Philip E. Steinberg is Professor of Political Geography and Director of IBRU, the Centre for Borders Research, Durham University, UK.

Jeremy Tasch is Associate Professor, Department of Geography and Environmental Planning, Towson University, Towson, USA.

Hannes Gerhardt is Associate Professor, Department of Geosciences, University of West Georgia, Carrollton, USA.

'*Contesting the Arctic* is one of the most significant recent works of Arctic scholarship ... The book gives readers unparalleled insight into the Arctic's current state and future ... the work is a must-read for those interested in the High North and international affairs generally.'
– *Foreign Affairs*

'A concise, inviting and crisply written analysis of the key narratives that inspire basically all decision-making in the Arctic ... *Contesting the Arctic* stands out as original and groundbreaking in the steady flow of new material on the Arctic.'
– *Arctic Journal*

'*Contesting the Arctic* is a sophisticated analysis of how contemporary discourses and performances are caught up in older colonial and Cold War legacies of knowledge production and geopolitics. It is a reminder to us all that we need to be ever vigilant in terms of how vast and complex spaces such as the "Arctic" are constituted and reproduced in political and popular cultures. As global attention grows towards the Arctic, this book reminds us that the Arctic is also a homeland and not an "empty space" to be scrambled over.'
– Klaus Dodds, Professor of Geopolitics, Royal Holloway, University of London

CONTESTING THE ARCTIC

POLITICS AND
IMAGINARIES IN THE
CIRCUMPOLAR NORTH

PHILIP E. STEINBERG, JEREMY TASCH AND HANNES GERHARDT
WITH ADAM KEUL AND ELIZABETH A. NYMAN

I.B. TAURIS

LONDON · NEW YORK

New paperback edition published in 2018 by
I.B.Tauris & Co. Ltd
London • New York
www.ibtauris.com

First published in hardback in 2015 by I.B.Tauris & Co. Ltd

ISBN: 978 1 78831 156 4
eISBN: 978 0 85773 844 8
ePDF: 978 0 85772 672 8

A full CIP record for this book is available from the British Library
A full CIP record is available from the Library of Congress

Library of Congress Catalog Card Number: available

Typeset in Garamond Three by OKS Prepress Services, Chennai, India

This book is dedicated to the over 200 individuals who took the time to speak with us in hopes of furthering open dialogue in the Arctic.

CONTENTS

ACKNOWLEDGEMENTS

This book has been a long time in the making – from 2007, when the initial project was conceived, through 2013 when the final manuscript was submitted – and over that time a large number of individuals has contributed. Our first debt of gratitude goes to the 'with's on the title page of this book: Elizabeth (Liz) Nyman and Adam Keul. As an UNCLOS expert in her own right who went on to serve as postdoctoral researcher for this project, Liz participated in the 2010 interviews in Washington and Ottawa, managed the transcription of interviews, and drafted Chapter 7. Adam came to the project after interviews were completed but went on to assist with transcriptions, do all the coding, and draft Chapter 5. This book would not be the same – and the writing of it would not have been so enjoyable – without their valued participation.

A host of other individuals has made significant contributions to the project, but three names in particular stand out. Rob Shields played an important role in conceiving and executing the first phase of this project, as well as authoring the Foreword. His insights from a long career investigating the meaning of 'the North' in southern (and, especially, Canadian) imaginations, coupled with his knowledge of the workings of Canadian government, were key in getting this project off the ground, and his participation in the 2008 Ottawa and Washington interviews provided invaluable assistance. Additionally, while we were just beginning to conceive this project, Klaus Dodds, an experienced Antarctic researcher who himself had recently begun to develop an Arctic focus, lent us his ear and his knowledge, and he has remained an informal advisor to the project throughout its duration. Ron Doel became involved with the project a bit later, and we are especially grateful to him for serving as external facilitator for the post-fieldwork debriefing meeting that resulted in the outline for this book. We look forward to many future collaborations, whether formal or informal, with all three of these stellar Arctic researchers.

Others who contributed to the project were Sandra Fabiano, who participated in 2008 Washington and Ottawa interviews, and Mauro Caraccioli, who participated in 2010 Washington and Anchorage interviews. Additional assistance with transcription was provided by Michael Husebo and Kelsey Scheitlin. Logistical support for fieldwork in Russia was provided by Vekoslav (Veko) Koshkin and Olga Tomolina. Other individuals who contributed to this project include Barret Weber, who shared his insights from fieldwork in Iqaluit and Nuuk, Anna Kerttula of the National Science Foundation, who has been a tireless advocate for this and other Arctic research projects, and David Stonestreet at I.B. Tauris, who maintained his confidence and patience throughout.

Funding was provided by the International Council for Canadian Studies, which supported the initial 2008 study on differing perceptions of the Northwest Passage; the US National Science Foundation's Geography & Spatial Science and Arctic Social Science programs (grants no. BCS-0921436, BCS-0921424 and BCS-0921704), which supported the 2010 interviews across the circumpolar Arctic as well as subsequent meetings of the researchers; and the European Commission's Marie Curie International Incoming Fellows program (grant no. IIF-GA-2010-275846), which enabled one of the authors (Steinberg) to devote parts of 2012 and 2013 to this manuscript as part of his Global Alternatives for an Interconnected Arctic project. We are also grateful to the journal *Politik* for permitting selected paragraphs from Hannes Gerhardt's article 'The Inuit and Sovereignty: The Case of the Inuit Circumpolar Conference and Greenland' to be reprinted in Chapters 4 and 6.

We also thank those who provided images for use in this book. We are especially grateful to Mike Avery, John K. Hall, Gita Ljubicic and Linda Scates for providing photographs from their personal collections. For all images, we have made efforts to identify and obtain permission from the rights holder.

Finally, we owe a debt of gratitude to the more than 200 individuals who spoke with us in individual and group interviews and trusted us with their stories, their disappointments and their dreams for the future. Without their cooperation, this project would not have been possible.

FOREWORD
BY ROB SHIELDS

Over the last few years, the position of the Arctic in the media and in the economies of the countries that ring it has begun to change, increasing the visibility and public awareness of Arctic issues. At the forefront of these issues is climate change. Where once the image of the Arctic was as fixed as the frozen hand of Franklin reaching for a Northwest Passage, it is now new and noteworthy again. Although the crystal clear distances of Greenland's fjords still afford the most spectacular visual experiences available on the surface of the planet, these once inaccessible, challenging places have become brochure destinations. These views are now a click away rather than requiring months of preparation, teamwork and diligent investment to experience and then return to tell the tale.

The Arctic has furnished dramatic images of expanses of icebergs suddenly breaking off from ancient glaciers, armchair specials on television and satellite views of fluctuating ice coverage on the Arctic Ocean. In recent years there has been a new spate of explorers from many countries. Russians travelled by two Finnish-built submersibles to plant their flag at the North Pole on the ocean floor in 2007. The Chinese steered a refurbished Russian icebreaker through the icepack to plant their flag at the pole in 2012. Local inhabitants of the Canadian Arctic islands have been surprised by tourists on cruise ships and yachting enthusiasts sailing through the unexpectedly ice-free narrows of the Northwest Passage. These adventurers have not been deterred by the gaps in technical images of the North, where reefs and depths are hardly plotted on incomplete navigation maps. The partial state of surveys and maps reminds us, every time a ship goes aground in the Baffin Strait, that the Arctic is still an area of white spaces on maps – of incomplete images, of both known and unknown unknowns. To fill in these white spaces, outsiders and Arctic residents alike rely on what the authors of this book term 'imaginaries'.

The renewed circulation of images of the Arctic – especially from above, below and along its more travelled coastlines – has unfortunately not triggered much reflection on their import. Over the last four years, the authors of this book have interviewed specialists and diplomats to capture the changing national viewpoints on the Arctic, and ventured north to grasp the impact of policies imposed from metropolitan centers. Knocking on the doors of institutes and ministries, embassies and think tanks, they found bureaucrats struggling to reconcile inherited policies with the changing realities of the North. Often respondents couched the policies they were designing, implementing, or interpreting the effects of through mytholo- gized notions of what past events meant for sovereignties, ignorant of what had actually taken place.

Past events become myths. Did an Inuit hunter brandishing a rifle on the ice really cause the *Manhattan,* an ice-strengthened tanker that challenged the Northwest Passage in 1969, to stop and thus become frozen in the ice? Are secret American submarine tracks across the Northwest Passage fully represented on the map over the door of the naval officers' club in Washington? And, is it true that Canadian representatives are routinely on board American nuclear submarines under Arctic ice – despite affronted Canadian posturing?

In sorting through these myths, the authors of this book detect competing 'imaginaries' that are hardly acknowledged, as competing voices speak past each other in attempts to impose their will and vision on the North and on competing countries' activities in the Circumpolar Arctic.

Imaginaries are not simply a matter of images, of superficial snapshots. Imaginaries are more importantly *imago,* foundational myths that provide a framework and reference for everyday life and for future ambitions. They are known through their effects. Even if ignored or denied, they are contours felt and grasped by busy hands of the present. They link the past to the now, just as the layered ice of glaciers shows us the planet's climatic memory, now set to 'Erase' in a slow melt as the climate warms. Likewise, today's images and views of the Arctic are changing with the warming of air and sea.

It is certainly a new thing to associate the Arctic with novelty and vanishing, fleeting images. However, the Arctic has always featured an extraordinary dynamism of change, of dramatically different seasons, of extraordinary cultural change by Inuit and other northern indigenes and residents who have reached to grasp the levers of political power and imagination to shape their destinies as planetary leaders. Imaginaries have a link to futures, and they shape the ambitions of both northern and other interested societies.

An Arctic-based thriller from 2006, *The Last Winter*, dramatically encapsulates the contestation of imaginaries that are discussed in this book. In the film, an American oil company is investigating the possibilities for exploiting major oil and gas deposits in Alaska's far northern wilderness, while independent researchers assess the environmental impacts of the proposed project and share their reports with distant government representatives, whose position on development *vis-à-vis* environmental protection is left undefined. The supervisor of the oil and gas exploration team clashes with the lead researcher, as their visions for the future of the Arctic seem irreconcilable. The film's plot twists through a series of fatalities that are variously explained. Perhaps the icy barrenness of the region's landscape, known to have ended the lives of the most experienced polar adventurers, is somehow responsible for the deaths. Scientific analyses suggest the possibility that noxious gas releases associated with permafrost melt are to blame. And finally, as unexplained apparitions are imagined, the more fanciful of the survivors' reflections propose that vengeful spirit-ancestors of indigenous peoples are protecting traditional northern territories from industrial exploitation.

The final scene segues to a bare hospital room where the lone surviving researcher, having recovered from her Arctic ordeal, walks away from the building as a newsperson delivers a television broadcast in an adjoining waiting room. The camera pans to a featureless sky and a puddle-strewn ground, while echoes of an Arctic wilderness are heard in the background. The film thus ends ambiguously, leaving the viewer unsure about how the churning imaginaries of a fragile yet unforgiving Arctic environment, a resource frontier ripe for exploitation and northern territorial access and control, will continue to interact, compete and find resolution. Which, of course, is where this book begins.

1.1   The Arctic Region.

CHAPTER 1

IMAGINING THE ARCTIC

In May 2008, representatives from the five countries bordering the Arctic Ocean met in Ilulissat, Greenland and signed a declaration that was notable in that it broke absolutely no new diplomatic or political ground. Indeed, that was the point of the declaration: to assert that there was a status quo, that it was functioning fine and that there was no need to change it.

The status quo that was reaffirmed at Ilulissat by the 'Arctic Five' (Canada, Denmark, Norway, Russia and the United States) was no less than the fundamental premise that underpins the political organization of the world. It is a status quo that has prevailed in Europe since at least the seventeenth century, and since then has spread through colonization and war to the rest of the world. Known as the modern state system or sometimes the Westphalian state system, in reference to the set of treaties that ended the Thirty Years War in 1648, this status quo is generally recognized as one that confers absolute sovereignty to the rulers of territorially defined states.

Although the most frequently noted aspect of the modern state system is that it divides land into state territories, it rests on an even more fundamental division of Earth's surface that makes the state-ideal possible: a division between land, which is deemed as having the potential for *territorialization* (that is, division into bounded and governed territories controlled by individual states) and water, which, except for internal waters like rivers, lakes and bays, is designated as beyond state territory. In recent decades, and especially since the United Nations Convention on the Law of the Sea (UNCLOS) was signed in 1982, individual states have come to receive certain rights in specific zones of the ocean adjacent to their coasts. States now have exclusive rights to resources (e.g., fisheries, offshore oil deposits) out to the 200 nautical mile limits of their exclusive economic zones (EEZs) and, if certain geological and bathymetric conditions are met, these rights may be extended for the seabed minerals of their outer continental shelves out to 350 nautical miles, or, in some cases, even beyond that. Notwithstanding these limited economic provisions, however, the *political* status of the ocean, beyond the 12 nautical mile coastal strip of territorial water, is high seas, a global commons outside the authority of any state. There are no provisions for coastal states to extend their limited economic rights in EEZs and outer continental shelves into anything that approaches the sovereignty that they claim on land. UNCLOS thus reaffirms a fundamental division of Earth's surface associated with the modern state system: sovereign land territories are juxtaposed against a global oceanic commons that is regulated *by* the state system but cannot become the territory of any individual state. Four of the five states present at the Ilulissat meeting have ratified UNCLOS and the fifth – the United States – while refusing formally to accede to the treaty, has declared that it endorses its fundamental principles as customary law.

The Ilulissat Declaration asserted that the modern organization of the world applies in the Arctic just as it does everywhere else (with, arguably, the partial exception of Antarctica). The signatories of the declaration acknowledged that the Arctic presents some unique challenges that may require extra levels of cooperation, and they expressed a desire to work with each other to address these challenges. In particular, they noted the value of the Arctic Council, a body whose working groups study and occasionally make policy recommendations regarding safety of navigation, search and rescue, environmental and development issues in the region and whose members include, in addition to the Arctic Five, the three other Arctic states (Finland, Iceland and Sweden) as well as six permanent participants representing indigenous peoples of the region. However, notwithstanding this recognition of the region's unique characteristics and challenges, the declaration makes it clear that land in the Arctic is clearly under the control of individual states, that the ocean is a global commons to be governed according to UNCLOS, and that institutions of cooperation will be tolerated only so long as they work within, and do not seek to override, this framework.

In short, the Ilulissat Declaration is not, in itself, a particularly interesting document, since all it does is assert that the Arctic is not exceptional. What *is* interesting, however, is that three foreign ministers (from Denmark, Norway and Russia), a Deputy Secretary of State (from the United States) and a Minister of Natural Resources (from Canada) felt that it needed to be written at all.

Presumably, the only reason to produce a declaration asserting that the Arctic is 'normal' would be if someone else were suggesting otherwise. But who was this, and what were they suggesting? No 'opposition' is specifically named in the declaration. Four weeks after the Ilulissat meeting, members of the research team behind this book began conducting a previously scheduled series of interviews on Arctic sovereignty issues, and at the last minute we added a question to the interview guide, asking respondents what they felt the declaration was in opposition *to*. Although the respondents were all knowledgeable observers and makers of Arctic policy, the answers to this question were surprisingly diverse. One US government official said,

> My guess is that there might be some green party in Europe that might be promoting [a treaty]. I heard some discussion of it in Scandinavia.... I've heard it mentioned in some European meetings, but I've never seen any substance.

Another American, a think-tank researcher with strong connections to the US military and foreign policy communities, asserted that the declaration was directed not at European environmentalists but at NATO, whom he was convinced had designs on the region. Still another policy professional, with a long history of involvement in international Arctic initiatives, speculated that, given Denmark's role as lead instigator behind the meeting and that it happened in Greenland, there might have been a subtext within Danish politics concerning Denmark's relations with its independence-seeking overseas territory.

The most complete explanation was given by an official with Foreign Affairs and International Trade Canada, who contextualized the May 2008 meeting within two high-profile events that had thrust the Arctic onto the world stage during the previous summer: the brief melting of sea-ice over the length of the Northwest Passage (the potential sea route through Canada's northern archipelago) and the planting of a titanium flag by a Russian submersible at the North Pole:

> Last October [2007], there was a meeting of legal advisers in Oslo, which included an American, a Canadian, and, I don't know who all was in the room. But it was certainly more than the five who were in Greenland. Then Denmark decided that they wanted to take this to the ministerial level. So they invited the five states which actually border on the Arctic Ocean, and they drafted a declaration which quite frankly said the same things that officials had already agreed to, which was that the Arctic is an ocean, and there is a Law of the Sea Convention which manages all aspects of the Arctic. There is simply no comparison between the Arctic and the Antarctic.... I think it was a very wise idea to get out ahead of the issue [by holding the Ilulissat meeting]. Just simply do a Google search on the Russian flag on the North Pole last summer, and you'll see in the first instance there's this enormous wave:

'Here they come, Russians over the top.' ...What was being done in Ilulissat is that the five countries at the level of their foreign minsters were standing up and saying, 'No, we've got a regime here, cool your jets.'... So essentially what [we were doing at Ilulissat] was getting out there and managing this issue. There's not going to be any new regime called the Arctic Littoral States or anything like that. The foreign ministers have put down their markers, everybody's signed on the dotted line, now we go back to work.

A similar explanation was given in 2010 by an official at the US State Department:

It would be a little bit of a misreading of the situation to say that we were enthusiastic about Ilulissat, because we were not. There was some reason to do Ilulissat, not the least of which was that it was coming on the heels of the Russian flag planting, when all the world's media were talking about an impending war in the Arctic because the Arctic countries were all racing to claim the shelf up there, and one of the purposes of Ilulissat was to show that that is not what was happening. The second purpose that you will see in the Ilulissat Declaration is to explain that the Law of the Sea already provides sort of a framework for the Arctic, [and so] there is no need for some sort of overarching treaty that is going to govern the Arctic like the Antarctic.

An official at the US embassy in Ottawa similarly stressed that Ilulissat was less in response to anything that was *actually* happening but to *perceptions* of what was happening:

Well I don't know that there was necessarily a specific proposal [against which Ilulissat was directed], but certainly in the media there has been a particular sense, and this goes back to global warming: 'What will climate change portend for the future? Is it going to be this race for the North Pole? Is there going to be this Wild West atmosphere where everybody is trying to sink as many wells as quickly as they can?' And then the Russians plant titanium-encapsulated flags, supposedly at the North Pole .... I think there is an awful lot of that in the public mind, and the media perpetuates this. It's a story, it sells a story, you put it on the front page: 'Canada reaffirms sovereignty in the face of Russian land-grab.' I think maybe, and as I say I don't know if there was necessarily a specific proposal informing anybody to do so, but even in some of the reporting in the local [Canadian] media in the run up to that meeting there was some discussion that the five countries were going to agree upon a new mechanism to control the Arctic. And from the paperwork that I saw, the documents that I saw ahead of time, that wasn't the idea at all. The idea was that they wanted the parties to reaffirm that in fact they would settle any differences responsibly.

As the Americans' explanations in particular make clear, the Ilulissat meeting was convened to develop a preemptive response to two, very different scenarios. Under one scenario the Arctic was emerging as a site where war could break out, as countries' efforts to map their outer continental shelf rights would devolve into an all out 'land-grab' (and 'sea-grab'). Although all individuals interviewed acknowledged that such a 'race for the Arctic' was not actually happening, they also recognized that the *perception* that such a race was occurring could lead to an actual race, and this could lead to general instability and conflict. Under the other scenario, which apparently was at least as frightening to representatives of sovereign states, perceptions that the 'race for the Arctic' was heating up could lend weight to calls for an Arctic treaty that would designate the Arctic Ocean as something other than an UNCLOS-governed global commons. Such a development would have the potential to undermine the near universal acceptance of UNCLOS, and theoretically it could even lead some to question core assumptions of the modern state system.

To be clear, the Arctic Five were not being irrational in issuing the declaration: one certainly could imagine a situation in which politicians in an Arctic nation were to rally their population to respond to what was perceived as another nation's incursion into 'our' Arctic, and this could bring about a spiral of militarization and counter-claims. One could also imagine that amidst such tensions (and perhaps outright conflict) others would respond by calling for a comprehensive treaty in order to 'restore the peace'. But nonetheless there is something extraordinary about sovereign states going out of their way to issue a declaration that asserts, in essence, that the Arctic is 'normal'.

Indeed, the very existence of a declaration affirming the Arctic's normalcy suggests that the Arctic is not completely normal. Or, if it is normal, there are enough perceptions of it being exceptional that its normalcy is contested, and therefore must be defended (and, in the process, potentially modified, whether ever so slightly as through institutions like the Arctic Council, or through more dramatic instruments). In short, preconceptions of what the Arctic is, and what it can be, matter profoundly. And this is all the more so as climate change transforms the Arctic's underlying character and as actors involved in making, interpreting, implementing, or responding to Arctic policies determine whether the Arctic is a 'normal' space that follows the standards of the rest of the world or is somehow 'exceptional'.

Imaginaries and Climate Change

A key theme in much of twentieth- and twenty-first-century social science has been to understand the links between how we experience the world, how we understand that world and how we order it (and thereby shape others' experiences and understandings). These systems of understanding and order extend to specific spaces: both the spaces that we encounter in our everyday lives and the spaces that we attempt to understand or control from afar. In the case of the Arctic, depending on one's perspective, the Arctic may be seen as an integral part of the existing nation-state, a sub-national indigenous group's homeland, a lost hearth of the national soul, a resource colony that is essentially empty of humans and that exists to be exploited (whether through mining or nature tourism) or preserved, a space of everyday activities (i.e., a 'home'), or the Arctic may be simply forgotten. As the circumstances surrounding the Ilulissat conference demonstrate, the Arctic may be seen as a 'normal' region or one that is 'exceptional'. If it is 'exceptional', this might be because of its geophysical characteristics, as a dynamic region of ice, land and water, or it may be due to its social characteristics, distant from southern capitals and populated by historically semi-nomadic peoples who cross state borders as well as externally based resource firms that have little interest in the region's long-term development. Or, most likely, ideas about the Arctic's exceptionalism and normalcy stem from a combination of conceptions about both its geophysical and geopolitical properties and potentialities. These constellations of ideas about what the Arctic is, and what it can be, are what in this book we are terming *imaginaries*.

Imaginaries are not stable. For centuries, Europeans from outside the Arctic saw the planet's northern periphery as a surface to cross, not a destination. The first European explorers to the North, like those throughout the Americas, sought neither territorial acquisition nor the establishment of permanent settlements. Rather, eighteenth- and nineteenth-century European explorers such as Vitus Bering, Sir John Franklin and Fridtjof Nansen were primarily driven by a desire to chart a sea route through the Arctic to the Orient. Similarly, acting on the notion that the Arctic is somehow an 'in-between' separating two destinations, a three-person Soviet aircrew flew an ANT-25 biplane during the summer of 1937 *over* the North Pole on a 7,100 mile world record setting non-stop flight between Moscow and southern California, and then back. This flight across the top of the world took place just three weeks after an earlier ANT-25 had also flown over the North Pole to reach the Pacific Northwest, to the mix of excitement and concern of Oregon residents, before the airplane's flight crew doubled back and landed in Vancouver, Canada. To the best of our knowledge, neither crew ever considered landing at the North Pole and exploring its icy environment.

More recently, policy makers who have attempted to direct attention to the Arctic have stressed that the region is not simply an empty space to be *crossed* by people from distant lands but that it is an arena of *connections* that can bring the world together. An early proponent of this view was the Canadian–American anthropologist Vilhjalmur Stefansson, who, in such books as *The Friendly Arctic* and *The Northward Course of Empire*, argued that the Arctic, far from being a frozen wasteland, was a 'Polar Mediterranean'. Like the Mediterranean, he wrote, the Arctic featured a relatively navigable central space that united diverse coastal peoples in commerce and productive interaction. As Stefansson wrote:

A map giving one view of the northern half of the northern world shows that the so-called Arctic Ocean is really a Mediterranean sea like those which separate Europe from Africa or North America from South America. Because of its smallness, we would do well to go back to an Elizabethan custom and call it not the Arctic Ocean but the Polar Sea or Polar Mediterranean. The map shows that most of the land in the world is in the Northern Hemisphere, that the Polar Sea is like a hub from which the continents radiate like the spokes of a wheel. The white patch shows that the part of the Polar Sea never yet navigated by ships is small when compared to the surrounding land masses.

This Mediterraneanist view of the region, which emphasizes proximate points on land rather than a separating ocean, was reproduced some 60 years later by then Soviet Premier Mikhail Gorbachev when he noted, in 1987:

The Arctic is not only the Arctic Ocean but also ... the place where the Eurasian, North American, and Asia Pacific regions meet, where the frontiers come close to one another and the interests of states ... cross.

Twenty years after Gorbachev, this theme was taken up again by former Alaska governor Sarah Palin, in what was probably the most quoted moment from her unsuccessful run for vice presidency of the United States:

We have that very narrow maritime border between the United States ... and Russia .... They're very, very important to us and they are our next door neighbor .... You can actually see Russia from land here in

Alaska, from an island in Alaska .... I'm giving you that perspective
of how small our world is and how important it is that we work with
our allies, to keep good relation[s] with all of these countries,
especially Russia.

While the Arctic imaginary held by explorers searching for the Northwest
Passage was of a hazardous region that, once conquered through its crossing,
could then be forgotten, the Mediterraneanist imaginary is, as Stefansson
would have it, a 'friendlier' space, one that can bring the world together. Still,
even this perspective is a view from the outside. In the Mediterraneanist
imaginary, the Arctic is seen as a space of *connection*, not *separation*, but it still
exists primarily to connect others who live *outside* the region. The Arctic is
therefore conceived as a land of proximate peripheries, not a center.
Individuals on the islands of Little Diomede (in Alaska) and Big Diomede
(two miles away, in Russia) may be able to wave at each other across
their fragment of the Arctic Mediterranean, but in the visions of both
Gorbachev and Palin they are located on peripheral extensions of countries
whose centers are very far away.

It should be no surprise that indigenous peoples of the region have their own
Arctic imaginaries, and these are quite different from those emanating from
southern capitals. This is not to say that there are 'pure' indigenous perspectives,
but rather to draw attention to the way in which indigenous peoples'
imaginaries flow from the ways in which their identities and livelihoods fit
within, or cut across, the norms established by the states in which they find
themselves, as well as the boundaries between land and water that underpin the
modern state system. And these imaginaries, like those of outsiders, are then
projected onto prospective political outcomes. These range from, for instance,
the formation of a new, postcolonial state in the Arctic (discussed in Chapter 4 of
this book) to innovative forms of socio-political organization that, by drawing
on traditional patterns of association and migration, transcend the modern
political organization of space (discussed in Chapter 6).

These imaginaries are themselves dynamic, and they have changed
considerably over time. The Cold War rivalry between the Soviet Union and
the Unites States, for instance, cut off most contact between Alaskan and
Russian Far Eastern indigenous peoples, and this in turn has led to differences
in cultural identities and political values. State assimilationist policies have
led to urbanization in some areas, while in other areas opportunities for work
in remote mines and oil fields have led to new migratory patterns amongst
populations that had been forcibly settled just a generation earlier.

In *The Last Imaginary Place*, Robert McGhee contrasts the idealization of
the Arctic as an 'imaginary' place – an isolated land that is frozen in time –

with the actual, dynamic Arctic that is characterized by connections and transformations, within and beyond its margins. While we join McGhee in rejecting the isolation narrative, we stress that imaginaries nonetheless play an important role as individuals and institutions, from both within and outside the Arctic, make sense of and respond to the region's dynamism. Furthermore, these intersections between geophysical changes, perceptions, mitigations and adaptations, and governance proposals – and between local dynamics and extra-local interactions – are particularly apparent with respect to climate change, which has had a particularly profound impact in polar environments. Although few of the greenhouse gases that are the primary contributors to climate change are emitted at the poles, an immediate impact of these emissions is a rise in temperature that has been twice as great there as in the rest of the world. In the Arctic, global climate change is leading reflective sea-ice to be replaced by heat absorbent water, which then creates a localized warming effect on top of that being experienced globally. Localized warming, in turn, leads to a decrease in land-based ice cover (as well as a further decrease in sea-ice), which leads to global sea-level rise. Amidst these warming trends, human and non-human life in the North struggles to adapt: hunting habitats disappear, coastal villages erode into the sea, thinning ice threatens traditional migration routes, melting permafrost causes roads to buckle and buildings to fall, and shortened ice-road seasons stymie industrial development. At the same time, these changes present new opportunities for mineral extraction, maritime transport, wildlife harvesting and even, on the region's southern periphery, agriculture. As Arctic landscapes and seascapes change, residents, states, corporations and others active in the North respond and adapt, but rarely without contestation and disagreement. In the process, all parties make recourse to imaginaries – ideas about what the Arctic is and about what it can, or should, be. Arctic imaginaries, like the Arctic itself, are never settled.

The Status Quo as Imaginary

To explore the dynamism and ambiguity of imaginaries in the Arctic, it is useful to return to the Ilulissat Declaration. The declaration was intended to reaffirm the status quo, reproducing an imaginary inherited from the modern system of sovereign states. And yet, according to many of its critics, the declaration, and, in particular, the process by which it was achieved, went well beyond the status quo, setting a new course for the Arctic, and one that could ultimately lead to the region being placed within a new imaginary that would sanction a higher level of appropriation and enclosure, whether by individual coastal states or by the Arctic Five acting as a collective entity that would exclude outsiders.

Not surprisingly, the most vocal opposition to the Ilulissat process came from those who were not invited to the meeting. By holding the meeting outside the context of the Arctic Council, the Arctic Five effectively excluded the three other Arctic Council member states, the six indigenous peoples' organizations that have permanent participant status in the Arctic Council and the various non-Arctic countries and non-governmental organizations that have Arctic Council observer status. The excluded parties' criticisms apparently were heard because, when a follow-up meeting of the Arctic Five was held two years later, in Chelsea, Québec, it fell apart in disarray amidst US Secretary of State Hillary Clinton's criticism of the process (and, in particular, the Canadian hosts) for their failure to provide a voice for indigenous peoples.

The same State Department official who defended US participation at Ilulissat explained what transpired at Chelsea:

> In terms of Chelsea, the last line of the Ilulissat Declaration makes it clear that the Arctic Council is the main place where we should be doing high-level diplomacy, [but] the idea of doing a second meeting of the Arctic Five started to imply that there was sort of an ongoing process . . . [Secretary Clinton's] overall remarks were talking about areas of cooperation, but again making the point that for diplomatic purposes, for high-level diplomatic purposes, for circum-Arctic issues, the Arctic Council should be the place where we are doing that. We already have this organization, we already have this forum where we have all the major stakeholders, or most of the major stakeholders, involved. One of the real concerns with the Arctic Five format was that there is no voice for the indigenous people at the table.

In fact, this is a selective telling of the Ilulissat Declaration, the final paragraph of which reads, in its entirety:

> The Arctic Council and other international fora, including the Barents Euro-Arctic Council, have already taken important steps on specific issues, for example with regard to safety of navigation, search and rescue, environmental monitoring and disaster response and scientific cooperation, which are relevant also to the Arctic Ocean. The five coastal states of the Arctic Ocean will continue to contribute actively to the work of the Arctic Council and other relevant international fora.

This would seem to fall short of the State Department official's claim that the Ilulissat Declaration reaffirmed that 'the Arctic Council is the main place where we should be doing high-level diplomacy'.

But even if the declaration did give primacy to the Arctic Council as the region's 'main' international forum, one must wonder why the Arctic Five chose to create a structure *outside* the Arctic Council in order to reaffirm the Arctic Council's centrality. A possible explanation was given by an official at the Canadian Embassy in Washington, who suggested that even as the Arctic Five countries sought to send a message that the Arctic was stable and under their leadership (with the Arctic Council serving as one of the pillars of that stability), they – and, in particular, the United States – were careful to avoid any move that might imply that the Arctic Council might become a *policy-making* body:

The Americans are very concerned about ideas, and as far as I know it's only ideas, to politicize the Arctic Council, to make the Arctic Council become more of a policy body and less of a scientific research body. I suspect Ilulissat sort of came out of [those concerns]. Well, we need[ed] to talk about these issues [but] the Americans don't want these issues talked about [in the Arctic Council]. So [they decided], 'Let's have a meeting. And maybe this is a place where we can talk about this stuff.'

In other words, the only way that the Arctic Five, and the United States in particular, could produce a statement that affirmed the role of the Arctic Council as supportive, but not generative, of policy was to hold a policy forum outside the Arctic Council framework that reaffirmed the importance of the Arctic Council. These legal machinations were not appreciated, however, by those who were not able to attend because the meeting was held outside the Arctic Council framework.

The message of Ilulissat was further muddled by the meeting's occurrence amidst intensified seabed-mapping. Indeed, it was seabed-mapping, and in particular the flag-planting at the North Pole that occurred as an ancillary to Russia's seabed-mapping project, that spurred the meeting in the first place. At Ilulissat, the five Arctic Ocean coastal states gathered to say, in effect, 'There is not a "scramble for the Arctic". We have the Arctic covered and we are expanding our influence in the area through an orderly process by which we are mapping our outer continental shelves in order to make the claims to seabed resource rights that are sanctioned under Article 76 of UNCLOS.' To the rest of the world, however, it appeared as though the Arctic Five were saying, 'We are abandoning the one forum that has been designed to give stakeholders who are not Arctic Ocean coastal states (e.g., indigenous peoples, the other three Arctic Council states, non-governmental organizations) a voice in the future of the Arctic commons and instead we are reasserting our right to 'claim' ever greater portions of the Arctic Ocean for ourselves, effectively crowding out all other actors.'

To be clear, the Arctic Five states' mapping projects, carried out for purposes specified in UNCLOS, were not designed to incorporate ever-increasing portions of the Arctic within sovereign state territory: they merely were to delimit the area in which each state would have sovereign rights to seabed minerals. This subtlety, however, was often lost on observers. For many, the Ilulissat Declaration, which was specifically designed to forestall the imaginary of the Arctic as a *terra nullius* ripe for claiming by adjacent states (an imaginary that is discussed in more detail in Chapter 2), inadvertently served to confirm it.

The story of the Ilulissat Declaration, the forces behind it, its intended meaning and its mixed reception reveal the contested nature behind even what is purported to be the status quo imaginary. If interested parties can have such varied interpretations of an imaginary that is intended, at root, to say 'The Arctic is a normal place', then how much more contestation can we expect around imaginaries that propose less 'mainstream' visions? And how much more contestation can we expect not just *within* imaginaries but *between* them?

Our point here, and a key point of this book, is that contestation among and within imaginaries is itself a normal state of affairs, in the Arctic as elsewhere. In this sense, the fact that there are competing visions for the region's future, based on competing ideas of what it is today, should not raise alarm. It does mean, however, that focused effort is needed to understand the mosaic of Arctic imaginaries.

Researching Arctic Imaginaries

In our effort to interpret the imaginaries that underlie various perspectives for the future of the Arctic, we were confronted early on with the apparent need to construct our *own* imaginary: What is the geographic scope of the Arctic? This is a question to which regional specialists have no definitive answer. A simple boundary would be the Arctic Circle (the latitude line at 66° 33' 44' North, north of which there is no direct light at the winter solstice). However, this also appears overly restrictive; the people of Nuuk, the capital of Greenland, and Iqaluit, the capital of Canada's majority Inuit territory of Nunavut, would likely be surprised to learn that they are not 'Arctic'.

Individual countries have different ways of defining 'The Arctic', or 'The North'. In Canada, the term usually (although not always) refers to the three northern territories, and in Norway the term 'High North' typically refers to the three northern counties, as well as the Svalbard island group. In the United States, many residents of the contiguous 48 states likely think of the whole of Alaska as 'The Arctic', a designation that probably would be contested by residents of Juneau, or even Fairbanks. In Russia, the divide between the North and 'the rest' is perhaps most unclear, in part because ideas of 'North' are often conflated with those of 'East'. Among the Arctic Five, only in Denmark, where there is a large distance between a clearly non-Arctic country and a clearly Arctic territory (Greenland) is it relatively simple to define where 'The Arctic' begins, although this, too, is becoming complicated as climate change expands the agricultural potential of southern Greenland.

Closely connected with this question is that of just what is an 'Arctic country'. Is an entire country Arctic just because it has a part that is Arctic (whatever precisely that means)? This question is particularly relevant for the United States and Denmark, where the country's Arctic fringe is geographically and culturally distant from the vast majority of the nation's citizens. Many Americans would probably be surprised if they were to read the United States' Arctic Region Policy's assertion that 'the United States is an Arctic nation.' Is a country Arctic because it is one of the eight members of the Arctic Council? Must it have coastline on the Arctic Ocean to be truly Arctic (a definition that would limit it to the five nations that attended the Ilulissat conference)? And, perhaps even more profoundly, does it even make sense, when one is seeking to define the Arctic, to begin by assuming that the Arctic is a world of *states*? Perhaps instead we should be asking when a person (or a community) is 'Arctic'. But in that case, must one be from an indigenous Arctic nation? Would the descendent of settlers count? What about a mine owner from outside the region who invests in northern development? Given the ongoing controversies regarding membership in the Arctic Council, its powers, and the privileged position given to the five Arctic Ocean coastal states at the Ilulissat meeting, these are not questions that researchers can answer without careful consideration.

When we set out to undertake this research we purposely refused to establish a preset definition of what (or where) the Arctic was. Our goal, after all, was (and remains) to understand how *others* imagine the Arctic. However, we did need to limit the field of individuals whom we were seeking to interview. In a perfect world, we would be able to document the Arctic imaginary of every individual who saw him or herself as 'Arctic' or who had thought about the region, but such a project would clearly be limitless and, in the end, pointless. Instead, we chose to restrict our investigation to individuals who consciously – in their jobs or in their spare time – were seeking to implement, influence, or shape the dialogue on the future of the region. Such individuals, in order to be effective, must reflect and articulate the views of a larger body of constituents (this could be the members of an indigenous group, the people who contribute funds to a non-governmental organization, or the co-workers at a government bureaucracy). Therefore, one would expect these individuals to be particularly forthcoming about the future that they envisioned for the Arctic, and why.

A more difficult set of choices concerned geographic scope, as we set out itineraries before going to the field. We made the (admittedly problematic) decision to limit our interviews to key cities in the five coastal states. In part, this decision was mandated by finances: conducting interviews in the other three Arctic Council states (Finland, Iceland and Sweden), let alone other states with Arctic interests (e.g., China, Japan, Korea), and conducting interviews in distant northern communities would tax our budget which, while generously provided for by a number of funding agencies, was not unlimited. It also would, in some instances, tax our language skills. We did, however, conduct interviews with several key representatives of indigenous groups in national and regional capitals and at indigenous people's gatherings (e.g., the 2010 meeting of the Inuit Circumpolar Council in Nuuk). Of course, our decision to conduct interviews in cities and conferences and not in villages meant that we heard only the views of leaders, not the rank-and-file. However, this was an omission that cut across the communities whose views we sought to document, not just indigenous communities. We have immense respect for anthropological researchers who imbed themselves in communities in order to document how understandings and norms are reproduced through everyday activities. That, however, was not the purpose of this study.

The interviews themselves were undertaken in two phases. The first phase, in 2008, focused specifically on the ongoing dispute over the legal status of the Northwest Passage, and, with just a few exceptions, these interviews took place in Ottawa or Washington. Notwithstanding the limited focus of this pilot stage, a number of themes that emerged from these interviews went on to inform the more comprehensive phase that followed, and data from the 2008 interviews have been fully incorporated into this study. In the second phase, conducted in 2010, interviews were conducted across the five states that border the Arctic Ocean. The bulk of these interviews took place in Anchorage, Copenhagen, Moscow, Nuuk, Oslo, Ottawa, St Petersburg, Toronto, Tromsø, and Washington.

Respondents included retired and current members of national and regional governmental agencies, militaries and parliaments; industry association representatives and corporate officials; researchers and research institution administrators; leaders and activists with indigenous peoples' groups; and officials with environmental and other non-governmental organizations. To comply with US federal research ethics guidelines, the identities of all individuals interviewed have been made anonymous to the greatest extent possible, and individuals were informed before they were interviewed that their identities would not be revealed. While our commitment to preserving respondent anonymity may make the reporting a bit awkward at points, our assurance to respondents that their identity would be obscured led in several instances to a greater degree of candor than we likely would have obtained if the respondent had spoken 'on the record'. Across the two phases, more than 150 interviews were conducted and all but a handful were recorded (with the consent of the person(s) being interviewed), transcribed, and coded.

Imaging Arctic Imaginaries

There are at least as many Arctic imaginaries as there are individuals whom we interviewed (well over 200, since several of the 150 interviews were conducted with multiple individuals). Making sense of these imaginaries proved to be a challenge.

Early on in the project, we realized that what we did *not* want to do was identify five 'national imaginaries'. Even if we could somehow decide the appropriate scale for the 'nation' that had the imaginary (Is the 'Alaskan' imaginary distinct from the 'United States' imaginary? Is the 'Inuit', or the 'Yup'ik', imaginary distinct from the 'Alaskan'?), such an approach would quickly devolve into a study of the different nations' policies. Policy is important, as are national differences in policy and in the underlying imaginaries that those policies reflect and reproduce. But a comparative country study would detract attention from those imaginaries that cross (or, in some cases, construct) national borders. Data-gathering was, for logistical purposes, country based and each author was given responsibility for gathering data from specific countries based on his regional and linguistic expertise. However, the intention was always to compile and share findings from individuals and institutions, across the various countries, in order to identify imaginaries that, in various forms, were prevalent among multiple populations in multiple countries.

Instead of identifying one national imaginary for each Arctic nation, we have identified six imaginaries that cross borders, in addition to the status quo imaginary exemplified by the Ilulissat Declaration. These six imaginaries, in turn, can be placed into two groups of three.

The imaginaries in the first group, detailed in Chapters 2 through 4, extend the basic system of bounded, sovereign, territorially defined states to the Arctic, albeit with specific modifications that account for the region's unique cultural and geophysical character. In the first of these imaginaries, discussed in Chapter 2, the Arctic is conceived as a *terra nullius*, an unclaimed but potentially claimable space beyond the normative regulations of international law, where individual states are free to exercise their expansionist tendency, whether claiming land, water, ice, or seabed. Although few, if any, states or non-state actors in the region directly advocate this ideal, this perspective on the Arctic – or the fear that *others* hold this perspective – lies just beneath the surface, informing not just media coverage but also the statements and actions of those responsible for making, interpreting and implementing Arctic policy.

The second imaginary, discussed in Chapter 3, is in a sense a moderated version of the first. In this imaginary, the underlying matter of the Arctic is not *irrelevant* (as it is in the first imaginary), but it is *different*. Therefore different norms and legal regimes are needed to govern this region where ice cover reworks the division between land and water that elsewhere orders the world. States thus are presented not with a *terra nullius* where there are *unlimited* opportunities for territorial expansion, but with a unique space of *new* and *different* opportunities.

The third imaginary considered is perhaps the most conventional. Here, although the Arctic may not present opportunities for new *levels* of territorial formation, it does present opportunities for replicating existing forms. This is seen in the various calls for indigenous, and, in particular Inuit and especially Greenlandic, statehood.

In Chapters 5 through 7, we turn from imaginaries that conceive of the Arctic as suitable for bounding by sovereign territorial states to ones that highlight processes and alliances that transcend boundaries. In Chapter 5, we discuss how the Arctic is imagined as a *resource frontier*, a trove of opportunities for states, corporations and individuals whose roots are elsewhere and who seek not incorporation of territory but extraction of riches. Examples from this chapter show corporations, states and a host of partners working together to facilitate economic activity and realize this Arctic imaginary.

Chapters 6 and 7 focus on groups whose imaginaries are, at best, loosely incorporated within the 'resource frontier' imaginary that structures so much of life in the Arctic. In Chapter 6, the emphasis is on indigenous organizations that, in contrast with those profiled in Chapter 4, seek not statehood but a kind of indigenous nationhood that transcends not just state boundaries but state ideals. Rather than reproducing the idea of statehood and transposing it to the context of indigenous peoples, this imaginary uses indigenous worldviews to challenge the fundamental assumptions behind the modern, territorial state.

Chapter 7 turns to an environmentalist imaginary that views the Arctic as a space whose nature is pristine but endangered, and that therefore should be governed according to an ethic that transcends the prerogative (and the developmentalist ideals) of the sovereign state. This chapter focuses specifically on environmental groups and the ways in which they articulate this vision with the world of states in which they operate.

A common theme throughout each of these chapters is that none of these imaginaries is as extreme (or as 'pure') as it first seems. To return to the example of the Ilulissat Declaration, this document was intended to reproduce the status quo, but its very existence actually revealed that the status quo was not stable. Indeed, in its reception amidst the context of ongoing state-sponsored seabed-mapping missions, the document inadvertently lent credence to the idea of a land-grab (or sea-grab), and thus it sowed further discord among Arctic actors. Similar contradictions and complications are revealed in the other imaginaries.

Thus, as we explore in Chapter 8, while the Arctic is 'contested', this 'contest' is not, for the most part, between *states*. States, in fact – and, in particular the five states that front the Arctic Ocean – are generally working cooperatively to implement a shared vision of the Arctic based on the continuation of the status quo imaginary. Rather the contestation is occurring within and between *imaginaries*. In this sense, the Arctic is anything but exceptional; arguably the entire world is characterized by such contested imaginaries.

At the same time, however, the contestation that is occurring among imaginaries is quantitatively, if not necessarily qualitatively, different in the Arctic than it is in other parts of the world. Where else are those who are attempting to design a region's future confronted with such an ambiguous materiality, a global warming ground zero, an indigenous homeland, a potential indigenous state, a nascent global governance regime, a coveted wildlife park, an oil bonanza frontier, and a throwback to colonial land-grabbing all in one package? The future of the Arctic, like that envisioned by the Arctic Five at Ilulissat, may end up reaffirming its 'normalcy'. But the road to that end will be anything but 'normal'.

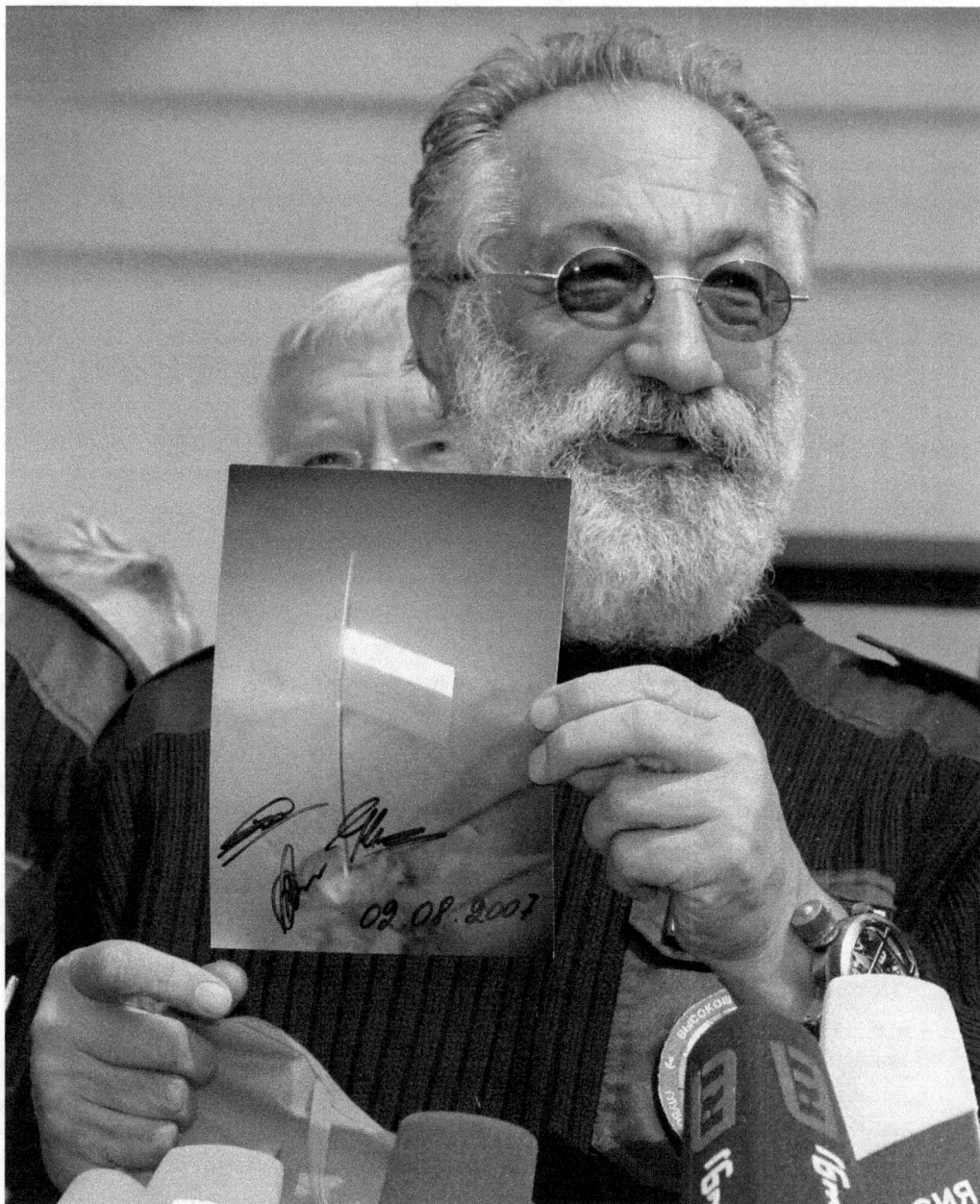

2.1 Artur Chilingarov displays a photograph of the Russian flag he planted on the seabed at the North Pole.

CHAPTER 2

TERRA NULLIUS

On 2 August 2007, the three-man crew of the submersible *Mir-1* laid down the gauntlet in a race for Arctic riches when they planted a titanium Russian flag on the seabed at the North Pole. Or, at least that's how the event was reported by the global news media. A photograph of the flag-planting was distributed worldwide, spurring commentators to liken the venture to earlier expeditions in which European countries claimed *terra nullius* – no man's land – by, quite literally, staking their claims. This parallel was explicitly referenced by Canada's Foreign Minister Peter MacKay in his rebuttal to the Russians: 'This isn't the fifteenth century. You can't go around the world and just plant flags and say, "We're claiming this territory".'

In fact, the message of the flag-planting was much more ambiguous. After all, the flag-planting led to the Ilulissat Declaration, whose signatories – including Russia – were intent on stating that there was *not* a race to claim Arctic territory and that the Arctic was *not* an unclaimed, but potentially claimable, *terra nullius*. Instead, the Ilulissat Declaration asserted that there was nothing left to be claimed in the Arctic: land was already securely incorporated as territory of individual sovereign states, while the ocean was affirmed to be an unclaimable commons, consensually governed by the global community of states according to the strictures of UNCLOS. But that leads us to ask a question similar to the one that we asked about the Ilulissat Declaration: Why did states make a big deal out of an event that they all were at pains to note was *not* a big deal? Put another way: what was it about the flag-planting that struck a nerve so deep that the Arctic Five felt that the Ilulissat meeting was necessary? The very fact that the Arctic Five felt that there was a need to affirm that the Arctic was 'normal' suggests that the flag-planting resonated with a persistent, if rogue, imaginary of the Arctic as somehow exempt from the modern division of the world, and that instead it was a region whose land and water were claimable by intrepid explorers on behalf of colonizing states.

Flagging Sovereignty

Much of the confusion surrounding the meaning of the Russian flag-planting stemmed from mixed messages given by different individuals associated with the mission. In a press conference shortly after the flag-planting, Foreign Minister Sergei Lavrov associated the flag-planting with Russia's effort to prove that the Lomonosov Ridge (and hence the seabed beneath the North Pole) was a geological extension of the Russian landmass. Russia was preparing to file a revised submission to the United Nations' Commission on the Limits of the Continental Shelf (CLCS) and, under the complicated rules for outer continental shelf claims elaborated in Article 76 of UNCLOS, if Russia could prove that the Lomonosov Ridge was an extension of the Russian landmass it could potentially claim exclusive rights to seabed minerals beyond the 350 nautical mile cut-off point that normally would apply.

Lavrov's explanation of the mission appeared straightforward but it allowed for varying interpretations regarding its ultimate significance. On the one hand, by associating the mission with efforts to prove contiguity between the North Pole and Russia's northern frontier, the foreign minister lent weight to comments like those of oceanographer and Russian Duma member Artur Chilingarov, who was a member of the expedition and who afterwards described it in tones suggestive of acquisitive colonization: 'The Arctic is Russian. We must prove the North Pole is an extension of the Russian landmass.' On the other hand, Lavrov's interpretation also put him in a position from which, when addressing international audiences, he could explicitly distance the flag-planting from colonial-style land-claims and instead place it within UNCLOS mandated (and therefore internationally sanctioned) scientific research: 'The aim of the expedition is not to stake Russia's claim but to show that our shelf reaches to the North Pole.'

Adding to the mixed messages emerging from Russia was Sergei Balyasnikov of Russia's Arctic and Antarctic Institute, who announced, 'For me this is like planting a flag on the moon.' On the one hand, this statement supported Lavrov's statement that the purpose of the flag planting was not to stake a territorial claim, since Balyasnikov's point was that Russia planted its flag on the seabed not to assert sovereignty (or even to stake out the area in which it should have sovereign rights to seabed resources) but to celebrate a technological achievement, just as was the case when the United States planted its flag on the moon. On the other hand, the equation of the North Pole with the remoteness and inaccessibility of the moon ran counter to the claim by both Lavrov and Chilingarov that the point of the mission was to demonstrate the North Pole's proximity to and contiguity with Russian territory.

Ambiguity concerning the message of the mission and its relation to whatever expansionist goals Russia may have had for the North was exacerbated by confusion regarding who actually was sponsoring the mission. Although the *Mir-1* and its companion submersible, the *Mir-2*, were in the Arctic assisting the research vessel *Akademik Federov* on a Russian government organized seabed-mapping expedition, the actual dive at the North Pole was organized by an Australian–American group of deep-sea exploration enthusiasts and funded by Swedish pharmaceutical magnate Frederik Paulsen. Indeed the combined crew for the two submersibles were, in addition to Chilingarov, Paulsen and the two pilots, Australian adventurer Mike McDowell and Russian businessman Vladimir Gruzdev. This is hardly the crew that one would expect if the primary purpose were to engage *either* in scientific exploration or imperial conquest. After the mission, McDowell explained that his group had sought Russian participation not to make a political statement but because Russia controlled two of the five submersibles capable of fulfilling the mission.

The Russians were certainly not alone in giving mixed messages regarding the intent and implications of the flag-planting. For instance, although, his 'This isn't the fifteenth century' statement implied that the Arctic was *not* a *terra nullius* suitable for claiming, Canada's Foreign Minister MacKay made other statements that suggested that the Arctic, including Arctic water, was indeed claimable (and, in fact, had already been claimed by Canada):

> [Russia] is posturing. This is the true North strong and free [a line from the Canadian national anthem], and they're fooling themselves if they think dropping a flag on the ocean floor is going to change anything . . . .

> We established a long time ago that these are Canadian waters and this is Canadian property . . . .

> The question of sovereignty of the Arctic is not a question. It's clear. It's our country. It's our property. It's our water . . . . The Arctic is Canadian.

Although Foreign Minister MacKay arguably took Russia's bait by responding to its 'claim' to Arctic territory with a 'counter claim', his statements were the exception. Throughout our interviews, respondents universally dismissed the idea that the flag-planting had any legal significance or that it in any way represented a Russian claim to sovereign territory. 'It was an amazing technological feat, but nothing else', remarked one US State Department official, while another American active in the Arctic research community noted:

> I think the media has sensationalized things, naturally, and it does capture people's imagination when you plant a flag somewhere. But I think from a governmental informed point of view planting flags went out of style as a way to claim territory about 400 or 500 years ago, maybe longer. . . . From a governmental perspective we don't get exercised by flags being planted along the Lomonosov Ridge.

Several respondents expressed their belief that the Russians had planted the flag for public relations purposes, both to increase enthusiasm for Arctic exploration (and military investment and resource extraction) among the Russian population and also to signal to the international community that Russia remained a committed player in the Arctic geopolitical arena. An official at the Norwegian Ministry of Defence likened the symbolic value of the flag-planting to Russian military flights over other Arctic countries' airspace:

> [These occur] because they still want to show that they're a great power. At least a great power if not a superpower anymore . . . . [The flag-planting] probably also [occurred] because Russia gets a lot of its income from the Arctic area in terms of natural resources, so they have to show the public that they will stake their claims in the Arctic. I think that has a lot to do with it as well.

The argument that the flag-planting was largely intended for a domestic constituency was reiterated by a Norwegian consular official in Russia:

> So Russia did place a flag on the seafloor, but this was only done for domestic consumption, to declare internally, so to speak, 'We are the champions.' But Russia's authorities fully accept the need to be accepted by international conventions.

An American active in the Arctic policy community made a similar point about how the Russian government was aware that the flag-planting had no legal significance, although, unlike the Norwegian, he stressed that the symbolic gesture was intended not so much for domestic consumption as for an international audience:

> It got global attention. *Time* magazine — a great land-grab and all that — write-ups in *Foreign Policy* . . . . It's international high-vis, but it has no relationship to international law, you know, and I think the Russian Foreign Ministry knows that and said that.

Of all the individuals with whom we spoke about the flag-planting, the only one who felt that it reflected an actual policy position, rather than a carefully constructed (or, perhaps, accidental) public relations effort, was an official at the Danish Ministry of Foreign Affairs:

> You will have noticed a sort of shift in the Russian position. They started out by planting a flag on the North Pole, saying, 'It's all ours.' Then they actually did participate in the Ilulissat Conference and agreed that the five Arctic states should cooperate under UNCLOS. And they are also sort of open to the eight Arctic states. They have moved all the way from the flag where they're saying, 'This is all ours,' to, 'Okay, we can work, we can be eight countries that work together on this.'

For this official, although the flag-planting did reflect an actual policy perspective, it was one that was soon abandoned.

If the flag-planting was just a public relations effort (or even if it were the expression of a rapidly abandoned policy), then it would seem that states and other Arctic actors could simply ignore the non-existent 'race' to claim *terra nullius* and instead address more pressing concerns. This was precisely the sentiment of a Russian indigenous activist: 'Putting the flag on the bottom of the North Pole was a PR event. But nonetheless industrial development will take place. This is reality. Our concern is how this can continue without interfering with native lifestyles.' And yet, the flag-planting, even if dismissed by some and ridiculed by others, put diplomats on high alert. A respondent at the US State Department noted that the event had done much more to raise the profile of Arctic policy among State Department officials than was the case in 2001, when Russia submitted its initial filing to the CLCS, even though she acknowledged that the formal claim to continental shelf resources had much greater legal significance. A researcher at a Norwegian defense policy institute similarly credited the flag-planting with awakening Norway's interest in the Arctic as a focus of national opportunity and potential insecurity. And, as we saw in Chapter 1, the flag-planting directly led to the Ilulissat Declaration, where the Arctic Five reasserted their understanding that, notwithstanding the impression given by the flag-planting that there was a 'scramble for the Arctic', the Arctic was a settled space, regulated by the norms of the modern state system.

The Persistence of Sectoral Thinking

In fact, while there is a consensus among policy makers that the modern state system, with its underlying division of claimable land from unclaimable ocean, extends to the Arctic, this has not always been the case. Beginning in 1907, in the absence of an international regime defining the status of the Arctic Ocean, Canada and then Russia contended that their borders should extend to the North Pole, via straight lines drawn from the easternmost and westernmost points of their national coastlines. This 'sectoral' division of the Arctic would grant each state a wedge of Arctic space that would enclose territory without any regard to whether it was land or water. Implicit in this sectoral claim was an understanding that the nature of the Arctic was so unique that it superseded the division between claimable land and unclaimable water that characterized the rest of the world. Subsequently, both countries abandoned their sectoral positions, and now, as in the Ilulissat Declaration, they restrict formal claims to *land* (and adjacent territorial and internal waters), while designating the Arctic Ocean itself as a high seas commons for free navigation.

Nonetheless, the sectoral principle, and the idea of claiming broad swathes of Arctic territory regardless of whether it is land or water, has continued to hold sway over the popular (and, occasionally, governmental) imaginations in several Arctic countries. Peter MacKay's statement that 'these are Canadian waters and this is Canadian property' and Artur Chilingarov's statement that 'the Arctic is Russian' were not sudden throwbacks to a policy that had been abandoned a hundred years earlier. Rather they emerged from, and resonated with, persistent mythologies of a national patrimony over Arctic land *and* water that have long been nurtured by some of the region's governments, even as these same governments have elected not to translate these claims into actual policy.

The case of Canada is particularly instructive here. Sixteen years before Foreign Minister MacKay reacted to the perceived threat from Russia's flag-planting by (unofficially) laying claim to Canada's entire Arctic sector, Belgian international law scholar Eric Franckx presciently wrote,

> [The sector] theory seems to exert a mystical attraction as a fall-back position whenever the Canadian sovereignty claim over its northern waters [has] to be buttressed .... It is obvious that for Canada the notion of [the] sector theory still has not totally fallen into oblivion.

Donald Rothwell, an Australian authority on international law in polar waters, has similarly written that, despite its official abandonment of any sectoral claim, Canada maintains a 'partial reliance upon the suspect sector theory'. Perhaps the most comprehensive analysis of Canada's tentative, but persistent, flirtation with making a sectoral claim has been produced by Canadian jurist Donat Pharand, who, in *Canada's Arctic Waters in International Law*, wrote:

The sector theory has been invoked by a number of politicians and officials in Canada as a legal basis for claiming jurisdiction not only over the islands of the Canadian Arctic Archipelago, but also over the waters within and north of the islands right up to the Pole. However, the government itself has never taken a very clear and consistent position on this theory. It would seem that present government policy is to hold the theory in reserve as possible support for its claim that the waters of the Archipelago are internal.

It should be noted that there is a significant difference between the claim that the waters of the Arctic archipelago (that is, the waters between Canada's Arctic islands) are internal to Canadian territory and the claim that a broad swathe of Arctic space up to the North Pole is Canadian. The claim to internal waters, which follows from Canada's announcement in 1985 that it was drawing straight baselines between the archipelago's islands, is based on Article 7 of UNCLOS, which specifies the conditions under which straight baselines (and the subsequent enclosure of internal waters) are permissible. Both the United States and the European Union have protested that the Canadian case does not meet these conditions, but, regardless of whether or not the Canadian interpretation of Article 7 is excessive, all would agree that it *is* based on an appeal to international law. In contrast, any assertion by Canada (or any other state) that it controls Arctic waters within its sector up to the North Pole would have no basis at all in international law.

In discussing Foreign Minister MacKay's comments after the Russian flag-planting, Canadian legal scholar Michael Byers has suggested that '[MacKay] got his speaking notes for the Northwest Passage [which runs through archipelagic waters] mixed up with his speaking notes for the Arctic Ocean.' That may well have been the case. However, if so, MacKay's slippage reflected a widely held conflation of the distinctions between archipelagic and far northern waters, and between the UNCLOS internal waters regime and claims to water territory within a national sector. These all get muddled amidst a general impression that the Arctic is 'different' and that its land *and its water* are integral components of the nation's patrimony, with the North Pole serving as a limiting point. In this way Foreign Minister MacKay was able to blend legalistic arguments about the internal water status of the Arctic archipelago with an emotional appeal to Canadians' natural rights in their polar sector, just as Foreign Minister Lavrov and Duma member Chilingarov were able to blend appeals to a Russian imaginary wherein Russian territory 'naturally' extends to the North Pole with a legally sanctioned process of scientific research mandated by UNCLOS. It also led Norwegian Minister of Petroleum and Energy Ola Borten Moe to proclaim in 2012, 'New areas will be opened up [for drilling]. There is no reason to stop now. Norway's present boundaries end almost right up at the North Pole,' even though, as Prime Minister Jens Stoltenberg reminded him in a rebuke the next day, Norwegian law prohibits drilling in far northern waters.

Geographer Klaus Dodds has suggested that the confusion, in both media reports and politicians' statements, between maritime sovereignty perform-ances that are clearly outside international law (which often reference the North Pole) and those that act to reaffirm international norms (which more typically reference UNCLOS-derived principles of internal waters, territorial seas, EEZs and outer continental shelves) stems from

> the uneasy co-existence between varied representations and understandings of the ocean including the seabed, ranging from the ocean as a procedural space to the ocean as material space, which is invested with resource potential, strategic access and cultural importance to coastal states and beyond.

Put another way, the sectoral imaginary of a claimed *terra nullius* is so powerful because it aligns with stories told within Arctic states about both their histories and their futures. And thus this imaginary persists, notwithstanding attempts by Arctic states to, officially at least, distance themselves from it.

While Foreign Minister MacKay's statements were unusually explicit in their reference to the sectoral ideal, similar thoughts prevail in other nations' ideologies, and in their attitudes toward the Arctic waters beyond their shores. For instance, in Norway, an official at the Ministry of Foreign Affairs explained the importance of 'Arcticness' in Norwegian identity in a manner that explicitly drew a swath of continuous Norwegian Arctic space, running from the coastline of northern Norway, through the offshore territory of Svalbard, and on through Arctic waters to the North Pole:

> We are living in the Arctic, we are part of the Arctic, and we have made use of the natural resources of the Arctic for centuries. So it's part of Norway, it's part of our identity .... We have an affinity to the Arctic that goes far back in time and at the same time we are part of the Arctic ....

> A lot of people [in Norway] historically got closely connected to or are related to people who grew up in the Arctic, either in the ocean, fishing, sealing, whaling, or on Svalbard in connection with coal mining. In northern Norway almost all small communities have a connection to

Svalbard because people went up there as miners. In addition, you've got the fishermen from the North, but also from the west coast of Norway, going up in the Arctic. They're fishing and still doing so. Norwegians identify themselves easily with the Arctic .... Everyone knows that Finnmark [the northernmost county of mainland Norway] is nearly all the way up to the North Pole via Svalbard.

Given these cultural attitudes, it is not surprising that Minister of Petroleum and Energy Ola Borten Moe made his misstatement about drilling in Norwegian waters 'almost right up at the North Pole'.

The promulgation of this imaginary that grounds ideas of national identity in Arctic sectors is not just innocently passed on through culture, however. It is also, in some cases, explicitly reproduced through government-sponsored public relations initiatives and educational tools. While Russia's campaign around the flag-planting is perhaps the most glaring example of a state promoting this imaginary, Dodds discusses how the Norwegian news release that accompanied the CLCS acceptance of Norway's filing featured a photograph of Norwegian Foreign Minister Jonas Gahr Støre pointing at the ocean north of Norway and how Canadian Prime Minister Stephen Harper has a penchant for posing for photographs in which he roots a foot in Arctic waters. With the promulgation of such images, Arctic states steer clear of making any formal assertions; images have no significance in international law. Nonetheless, such images serve to reproduce ideas among the populace about what is 'our' Arctic and what is 'our' identity as an Arctic nation.

A particularly sustained semi-official expression of the sectoral perspective can be seen in a series of maps produced by the Government of Canada. *The Atlas of Canada* (produced by Natural Resources Canada) contains a map called 'The Territories' that features two sectoral lines, one from the northernmost point on the Yukon–Alaska border to the North Pole, and the other from the northernmost point in the channel between Ellesmere Island and Greenland to the North Pole. Lest there be any confusion about the meaning of these lines, the legend informs the reader that they are 'international boundaries'. Even more provocative is the atlas's polar-projection map, 'North Circumpolar Region'. This map depicts these same sectoral lines (again designating them as 'international boundaries') but conspicuously omits other states' equivalent sectoral boundaries (i.e., those between the United States and Russia, between Russia and Norway, and between Norway and Denmark/Greenland).

During an interview with an official at Transport Canada, the conversation inadvertently turned to these maps. The interview was being conducted in

North Pole ★ Pôle nord

ARCTIC / OCEAN
OCÉAN / ARCTIQUE

CANADA

www.atlas.gc.ca

North Magnetic Pole
Pôle nord magnétique

ARCTIC OCEAN
OCÉAN ARCTIQUE

Lincoln Sea
Mer de Lincoln

Ellesmere Island
Alert
Île d'Ellesmere

KALAALLIT NUNAAT
(GRØNLAND)
(Denmark / Danemark)

Axel Heiberg I

Grise Fiord

Beaufort Sea
Mer de Beaufort
Banks I
Sachs Harbour
Melville I
M'Clure Str
Resolute
Devon I
Lancaster Sd
Baffin Bay
Baie de Baffin

Arctic Circle
Cercle arctique

ALASKA
Old Crow
Tuktoyaktuk
Inuvik
Fort McPherson
Paulatuk
Amundsen Gulf
Ulukhaktok
Nanisivik
Arctic Bay
Pond Inlet
Clyde River
Détroit de Davis
Davis Strait

USA / É-U d'A
Dawson
NORTHWEST TERRITORIES
Norman Wells
Cambridge Bay
Victoria Island
NUNAVUT
Baffin Island
Qikiqtarjuaq
Île de Baffin
Pangnirtung

YUKON
Mt Logan
5959 m
Faro
Déline
Great Bear
Kugluktuk
Gjoa Haven
Taloyoak
Igloolik
Hall Beach

Whitehorse
Watson Lake
Wrigley
Grand lac de l'Ours
Back R
Kugaaruk
Repulse Bay
Foxe Basin
Cape Dorset
Iqaluit
Kimmirut

BRITISH COLUMBIA
COLOMBIE-BRITANNIQUE
Fort Simpson
Fort Liard
Great Slave
Behchokò
Yellowknife
Lutselk'e
Thelon R
Baker Lake
Coral Harbour
Hudson Str Dét d'Hudson
NFLD & LAB
T-N-et-LAB

Hay River
Fort Resolution
Grand lac des Esclaves
Fort Smith
Rankin Inlet
Whale Cove
Arviat
Chesterfield Inlet

ALBERTA
SASK
MANITOBA
Hudson Bay
Baie d'Hudson

Sanikiluaq

QUEBEC
QUÉBEC
Ungava Bay
Baie d'Ungava

TERRITOIRES DU NORD-OUEST

Mackenzie R

LEGEND / LÉGENDE

○ Territorial capital /
Capitale territoriale

● Other populated places /
Autres lieux habités

—·—· International boundary /
Frontière internationale

—··—·· Territorial boundary /
Limite territoriale

— — Dividing line /
Ligne de séparation
(Canada and/et Kalaallit Nunaat)

Scale / Échelle
300   0   300   600   900
km   km

ONTARIO
James Bay
Baie James

N

© 2006. Her Majesty the Queen in Right of Canada, Natural Resources Canada.
Sa Majesté la Reine du chef du Canada, Ressources naturelles Canada.

2.2   The Territories, *Atlas of Canada*.

the Transport Canada office in Ottawa and, over the course of the interview, both the interviewers and the Transport Canada official began pointing at a version of the 'North Circumpolar Region' map that was tacked on the office wall to refer to specific places. This led the official to interrupt himself mid-sentence and interject:

Respondent: That map is inflammatory, by the way.

Interviewer: Because of the Canadian lines?

Respondent: Yeah … We know that it's wrong. The map makers should have known better, and interestingly this map was made for our National Defence people.

@ 2001. Her Majesty the Queen in Right of Canada, Natural Resources Canada. / Sa Majesté la Reine du chef du Canada, Ressources naturelles Canada.

2.3   North Circumpolar Region, *Atlas of Canada*.

Interviewer: I've often wondered: why doesn't Foreign Affairs complain about these maps? Unless everyone knows that they're unofficial.

Respondent: Everyone doesn't know it; nobody knows it . . . . And we have done it again recently on something called the Northern Strategy. We pointed it out to them, but they went ahead with it.

The official here was referring to *Canada's Northern Strategy: Our North, Our Heritage, Our Future*, a major policy document from 2009 that details international as well as domestic commitments and that is referred to frequently by Prime Minister Harper. The document contains two maps that include sectoral lines: an adaptation of the *Atlas of Canada* 'Northern Territories' map, and a reprint of a polar projection map that had been produced to celebrate Canada's participation in the International Polar

Year. The polar projection map, in which Canada – and only Canada – claims sectoral sovereignty, is placed, without an apparent recognition of irony, amidst text celebrating Canada's commitment to cooperatively engage its Arctic neighbors.

The persistence of sectoral signifiers is not just for public relations purposes: it also serves to keep alive an imagery that is sometimes deployed in boundary disputes. In the Beaufort Sea, where the one remaining maritime boundary dispute in the Arctic region persists, Canada argues that the line dividing its waters from US waters should be drawn according to an 1825 treaty between Great Britain and Russia that sets the boundary between what is now the Yukon Territory and Alaska at the 141st meridian 'in its prolongation as far as the frozen ocean'. Canada asserts that this language extends the meridian into the ocean, creating, in effect, a sectoral boundary line that, while still not going all the way to the North Pole, does go as far as the 200 nautical mile limit of the two countries' EEZs, and potentially further to demarcate the boundary between the two countries' outer continental shelves. In the Barents Sea, Russia similarly, until 2010, relied on a Soviet law from 1926 to claim that the maritime boundary between Russian and Norwegian waters was a sectoral line that ran close to the 32nd meridian. In both seas, the opposing state – the United States in the Beaufort, Norway in the Barents – has argued that such appeals to a sectoral division of the Arctic Ocean have no basis in international law and that instead these maritime boundaries, like others around the world, should follow the principle of equidistance detailed in UNCLOS, whereby boundary lines are drawn perpendicular to the coast. The Barents Sea dispute was settled in 2010 through a treaty that recognizes the principle of equidistance, but that then adjusts it in such a way that the previously disputed area between Russia and Norway is split roughly in half. The dispute in the Beaufort Sea remains.

Yet another reason for the persistence of the sectoral imaginary, even amidst its continual disavowal, is that it provides a convenient and legible means for geographically dividing space, especially for managing phenomena that *do* transcend the land–sea divide. Thus, for instance, when in 2007 the METNAV system – a collaborative venture of the International Hydrographic Office, the World Meteorological Organization and the International Maritime Organization that monitors meteorological conditions for navigators – expanded its coverage to the Arctic in anticipation of increased Arctic shipping, it divided the Arctic Ocean into five sectors defined by meridian lines that extend to the North Pole. With two Russian sectors, two Canadian sectors and one Norwegian sector (and no Danish or American sectors), however, these sectors were so clearly disassociated from the borders that would emerge if Arctic states were simply to extend their outermost boundaries toward the North Pole that there was little concern that they could be confused for territorial sea claims.

More problematic was the Agreement on Cooperation on Aeronautical and Maritime Search and Rescue in the Arctic (the Search and Rescue Agreement), agreed to at the 2011 Arctic Council ministerial meeting. This agreement was the first binding agreement negotiated under the auspices of the Arctic Council, and thus it attracted a high level of attention from those who were concerned about (or who celebrated) the Arctic Council taking a more proactive role in Arctic governance. Given the agreement's high political profile, as well as the ongoing discourse of a 'scramble for the Arctic', it is notable that the portions of the Arctic assigned to each of the Arctic Five states (and Iceland) for search and rescue purposes closely resemble those that would be drawn if the negotiators of the agreement had simply drawn sectoral lines from each state's outermost borders. Perhaps for this reason, Article 3.2 of the Search and Rescue Agreement contains the following disclaimer: 'The delimitation of search and rescue regions is not related to and shall not prejudice the delimitation of any boundary between States or their sovereignty, sovereign rights or jurisdiction.' As in the Ilulissat Declaration, the 'normalcy' of the Arctic is affirmed. But, again paralleling the Ilulissat Declaration, the very fact that state parties felt a need to make this affirmation suggests that there were underlying counter-currents of exceptionalism.

Images and Imaginaries

Of all the imaginaries discussed in this book, the imaginary of the Arctic as *terra nullius* may be the most nebulous. No one is actually advocating that the Arctic is *terra nullius*, and when a global policy-making body produces something that might imply the *terra nullius* imaginary – as in the Search and Rescue Agreement – special efforts are made to assert that the Arctic is *not* claimable or divisible according to any principles other than those that operate in the rest of the world.

And yet, we have also seen the power of *images* that portray the Arctic as an empty space that lies beyond the rules and categories of the state system and that in turn can be divided into wedge-shaped sectors of territory without regard to the particularities of land, sea and ice. A range of actors in the Arctic has deployed such images to suggest the *terra nullius* imaginary. By working with images that are open to multiple interpretations (with varying levels of official sanction), they are able to promote the imaginary contained in these images without actually advocating that the imaginary be translated into policy proposals.

Failure to translate the *terra nullius* imaginary into policy does not, however, make this imaginary or its images irrelevant. In one respect, as we have seen at Ilulissat and elsewhere, the prevalence of images that suggest the *terra nullius* imaginary has led to reactive policies that are promulgated for the main purpose of *refuting* the imaginary. The imaginary is also important because it has shaped how Arctic events are covered in the news media, which in turn impacts policy. Consider, for instance, the coverage of the various efforts that have been made by Arctic states to map their outer continental shelves in preparation for making claims with the CLCS. Across a range of media, these efforts have been highlighted as illustrative of a 'race for the Arctic'. States are seen as rushing about, seizing territory. Whether they are conquering and incorporating what previously was a global commons or, even more aggressively, taking territory from another state, when one views the Arctic from this perspective it is seen as a zone of likely conflict.

In fact, the seabed-mapping missions, although often portrayed as a competitive mad dash to gain sovereign territory, have been neither 'competitive' nor a 'mad dash'. Nor, in fact, have they been about 'gaining sovereign territory'. The entire process has been exceptionally orderly, as all states, in their *official* behavior (if not in their maps, unofficial statements and public relations campaigns), have followed the process laid out in Article 76 of UNCLOS. Furthermore, the purpose of these efforts has not been to add sovereign territory (which could potentially be translated into military power) but rather to define the areas of the seabed beyond 200 nautical miles from shore where they can claim exclusive resource rights. In technical terms, the coastal states are defining areas in which they have 'sovereign rights', not 'sovereignty', and they are marking 'limits', not 'boundaries'. And thirdly, despite its inherently competitive nature, the seabed-mapping process has actually been characterized by considerable cooperation, even among states that one would expect to be antagonistic toward each other. Notwithstanding their dispute over the Beaufort Sea, the United States and Canada have for several years engaged in joint seabed-mapping exercises, with the US contributing its expertise in gathering and analyzing bathymetric data and Canada gathering and sharing complementary seismic data, both of which are needed for filing claims with the CLCS. Similarly, Denmark and Canada have engaged in joint mapping exercises, notwithstanding their dispute over Hans Island, which is located between Greenland and Canada's Ellesmere Island and which remains the one disputed piece of land in the Arctic. Scientific institutes in all five Arctic Ocean coastal countries, plus institutes in Germany, Iceland, Italy, Spain, and Sweden, have been contributing Arctic bathymetric data to the International Bathymetric Chart of the Arctic Ocean (IBCAO), a collaborative effort hosted by the United States' National Oceanic and Atmospheric Administration. IBCAO data are made publicly available and can be used by any country to support its CLCS filing.

Beyond the example of seabed-mapping, similar stories of international cooperation emerged in many interviews. Individuals involved in activities ranging from hazard planning and response to defense and security, from wildlife management to the promotion of scientific research, and from navigational assistance to environmental monitoring, all stressed that not only was the Arctic not a zone of exceptional conflict: it was a zone of exceptional cooperation, both in data-sharing and in joint operations.

To some extent, misconceptions about just what is being mapped (and claimed) on the outer continental shelf stem from the sheer complexity of the legal formula used to define a country's limits. Outer limits for a country's claims are established by one of two lines: one that is located 350 nautical miles from the country's coast (or baseline) or a second line that is located 100 nautical miles from the farthest seaward point of the continental landmass at which the ocean is fewer than 2,500 meters deep. A country can choose whichever of these two measures gives it the greatest claim but, even then, this line marks only the outermost limits of a potential claim. Once the potential outermost limit is established, the country must choose one of two possible means for determining precisely how far one's claim can reach within that limit, either by drawing a line until it reaches the point 60 nautical miles seaward of the foot of the continental slope, or until it reaches the point at which sediment thickness is 1 percent of the distance to the foot of the continental slope. Needless to say, the complexities of this formula, the scientific research that must be completed before one can make a filing, and the specification of precisely what rights are granted to states once their filings are approved by the CLCS all tax the limits of what can be communicated in a newspaper article directed toward the general public. It is much easier to report on 'efforts being made to claim a portion of the Arctic Ocean', or better yet, to print an image of that 'claim' being performed, through seemingly 'official' acts of what Klaus Dodds calls 'flag planting and finger pointing'.

As Dodds stresses, these acts of 'flag planting and finger pointing' are neither accidental nor incidental to how we imagine the Arctic (and contest its future). Images matter. Indeed images may matter more in the Arctic than elsewhere because so many of the individuals and institutions who make policy for the region (or who elect these individuals or invest in these institutions) have no direct experience of it. In addition, in the Arctic sovereignty is often performed through images because the actual power of the state to control Arctic territory is muted.

In such a situation, the image itself takes on a life of its own, which is frequently exploited to support an imaginary. Thus, the image of a planted flag, or a map that claims swaths of space, is extracted from its original context and used to 'prove' that the Arctic – or at least a portion of the Arctic – is 'ours', even though this is in contravention of official government policy. The emotive power, but lack of legal standing, of images, allows governments to have it both ways: local populations can be rallied while the diplomatic community is assured that there is no intent to violate international law.

The problem is that the passions aroused by such images can get out of hand and overwhelm their legal insignificance, particularly when the images are received within the context of a broader imaginary of the Arctic as a claimable *terra nullius* as well as mythologies that define the Arctic as essential to national identity (as one finds, in varying degrees, in Canada, Norway and Russia in particular). If 'we' are an Arctic people and the Arctic is a place in which other nations are staking claims, then 'we' must defend 'our territory' (or, in Canadian terms, 'our sovereignty') in this 'race'. In the process of making such arguments, images of – and claims to – Arctic territories as stretching to the North Pole and existing in disregard of prevailing legal norms become increasingly acceptable.

One of the most frequently reproduced contemporary maps of the Arctic is 'Maritime Jurisdiction and Boundaries in the Arctic Region', published in 2008 (and subsequently updated) by the International Boundaries Research Unit (IBRU) at Durham University.[1] Through representing the Arctic Ocean in a mosaic of colors and hash marks, the IBRU map depicts the agreed upon (or, in some cases, disputed) maritime boundaries between Arctic states, their EEZs and the potential limits of their outer continental shelf claims (i.e., the 350 nautical mile and the 2,500 meters deep +100 nautical mile lines). The resulting map is of an Arctic Ocean that appears to be almost entirely divided among the coastal states.

The day after its 5 August 2008 release, the IBRU map was discussed extensively in 'Ocean Law Daily', an e-mail briefing distributed by Caitlyn Antrim, a former member of the United States UNCLOS negotiating team who went on to become Executive Director of the Rule of Law Committee for the Oceans, an organization that promotes US accession to UNCLOS. In the briefing, Antrim notes that her 'overall assessment of the IBRU chart is that it is designed [to highlight] the worst-case view of the status and prospects for Arctic claims and counter-claims.' In part, Antrim faults IBRU's decision to depict the two lines that show the outermost limits of potential claims (the 350 nautical mile and the 2,500 meter +100 nautical mile lines) but not the continental slope and sedimentary depth data that will, in the end, significantly reduce the extent of those limits.

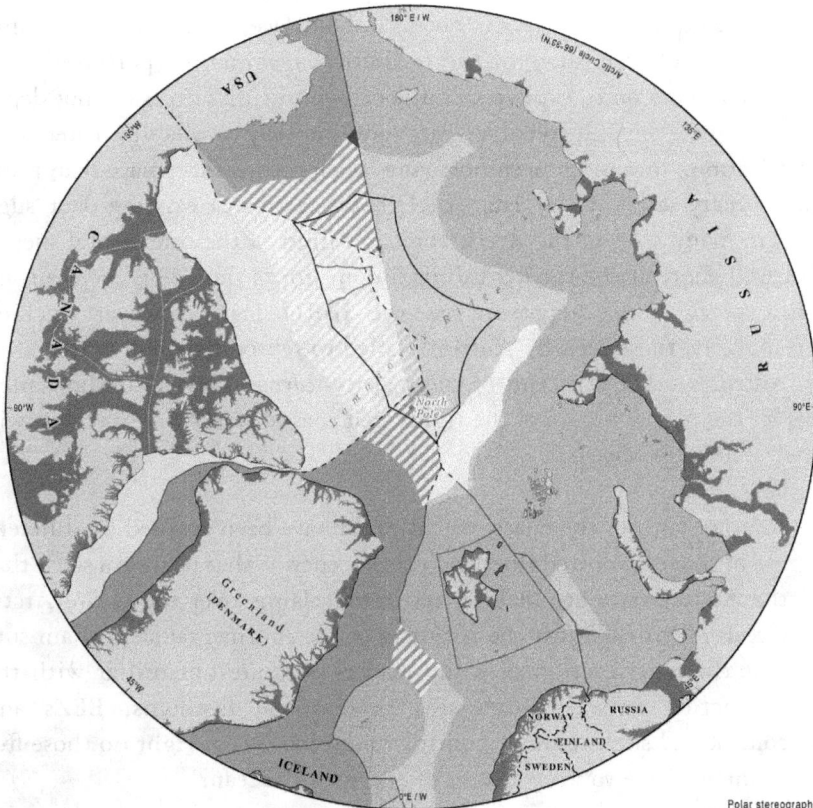

Polar stereographic projection

0	nautical miles	400 at 66°N
0	kilometres	600

Map labels: 160° E / W · Arctic Circle (66°33'N) · USA · CANADA · 135° W · 135° E · RUSSIA · 90° W · 90° E · 45° W · 45° E · North Pole · Lomonosov R · Greenland (DENMARK) · ICELAND · NORWAY · FINLAND · SWEDEN · RUSSIA · 0° E / W

Legend (left column):

- Internal waters
- Canada territorial sea and exclusive economic zone (EEZ)
- Potential Canada continental shelf beyond 200 nm (see note 1)
- Denmark territorial sea and EEZ
- Denmark claimed continental shelf beyond 200 nm (note 2)
- Potential Denmark continental shelf beyond 200 nm (note 1)
- Iceland EEZ
- Iceland claimed continental shelf beyond 200 nm (note 2)
- Norway territorial sea and EEZ / Fishery zone (Jan Mayen) / Fishery protection zone (Svalbard)

Legend (middle column):

- Norway claimed continental shelf beyond 200 nm (note 3)
- Russia territorial sea and EEZ
- Russia claimed continental shelf beyond 200 nm (note 4)
- Norway-Russia Special Area (note 5)
- USA territorial sea and EEZ
- Potential USA continental shelf beyond 200 nm (note 1)
- Overlapping Canada / USA EEZ (note 6)
- Eastern Special Area (note 7)
- Unclaimed or unclaimable continental shelf (note 1)

Legend (right column):

- —— Straight baselines
- —— Agreed boundary
- – – – – Median line
- —— 350 nm from baselines (note 1)
- —— 100 nm from 2500 m isobath (beyond 350 nm from baselines) (note 1)
- —— Svalbard treaty area (note 8)

2.4   Maritime Jurisdiction and Boundaries in the Arctic Region, IBRU, Durham University.

IBRU seems to have anticipated this critique. In the first footnote of the three-page publication, IBRU notes, in bold face, 'In reality, the claimable areas may fall well short of the theoretical maximums [portrayed on this map]' and goes on to explain that the other limiting factors are not depicted because not enough reliable data are publicly available. Later in the publication, in a note accompanying a supplementary map that presents preliminary seabed relief data, IBRU notes again that existing data 'suggest that in many areas of the Arctic the outer limit of the continental shelf may fall well short of the theoretical maximum limits shown on the main map'. Most news stories reporting on the IBRU map, however (including, presumably, those seen by Antrim), failed to reproduce these notes.

Antrim's other criticism of the map concerned how its design failed to depict the partial nature of the rights that coastal states would obtain within their delimited zones:

> Another fault of the chart, which could have been avoided by different use of shading, dotted lines and translucency, is that it gives a sense that the Arctic states are making territorial claims over the entire Arctic Ocean rather than just the resources of the continental shelf. I am sure that the first reaction of many readers who are unfamiliar with the distinction between high seas navigational freedoms, EEZs and continental shelf resource control would be: 'What right do those five countries have to keep us out of the Arctic Ocean?'

In the end, the combination of lines that map the 'worst-case' limits of states' outer continental shelf claims and the depiction of these limits in ways that imply that they signify the potential extent of state territory, left Antrim wondering if these aspects of the map were truly accidental. She concludes her briefing by speculating that this apparent effort to portray the Arctic as under siege by the encroaching five coastal states may be part of a scare campaign by the United Kingdom (and the European Union) to support their argument that special safeguards are needed to preserve access by non-coastal states.

Coincidentally, just two months after the IBRU map was released, the US State Department published its own map, 'The Arctic Region', and the differences between it and the IBRU map are striking. Although maritime boundaries (including some contested ones) are depicted on the State Department map, the ocean itself is all the same color, with some slight differences in intensity to indicate the extent of ice cover. Although the 200 nautical mile limit of EEZs is drawn on the map, there are no references to continental shelf claims or their potential limits.

The State Department map was conceived shortly after the Russian flag-planting, as one component of a series of efforts by the State Department and other nations' foreign ministries to demonstrate that the Arctic was an ordered ocean in which the rule of law prevailed. This was essentially the same message that was articulated by the foreign ministries at Ilulissat. A State Department official elaborated on how the map came about and what was behind the decisions to highlight some features and omit others:

> Before public dissemination [the map] was circulated within the Government. That map itself was in the works for almost a year and represents the collaborative efforts of [several offices at the State Department] and talented cartographers from several intelligence agencies. The lines were very carefully massaged to reflect US boundary depiction policies . . . .

> We carefully avoided consideration of any hypothetical claims . . . . There are powerful US politicians opposed to our acceding to UNCLOS and who go apoplectic about 'Russia grabbing the Arctic Ocean and its resources with UN approval.' . . . Thus we did not want to speculate on our map who might assert what entitlements to the shelf beyond 200 nautical miles while submissions to the Commission are still pending for the other Arctic States. . . . We must (and I believe, should) err on the side of caution in dividing up the Arctic Ocean.

In other words, just as the Canadian maps were designed to suggest that a portion of the Arctic Ocean was (or could be, or should be) Canadian, the US map was made specifically to give the message that the Arctic was, and would remain, no one's: not a *terra nullius* (an unclaimed space capable of being claimed) but a *res extra commercium* (a space incapable of being possessed).

The Strength of Titanium

The lesson behind all of these duelling maps – the *National Atlas of Canada* maps, the IBRU map, and the US State Department map – is that the sectoral approach, and the associated *terra nullius* imaginary, are still very much alive, in images that subtly (or not so subtly) promote it and in others that are produced and distributed with the specific intent of *denying* that any claims based on a *terra nullius* imaginary are possible. Contestation surrounding the *terra nullius* imaginary is amplified because the images that promulgate this imaginary themselves have multiple meanings.

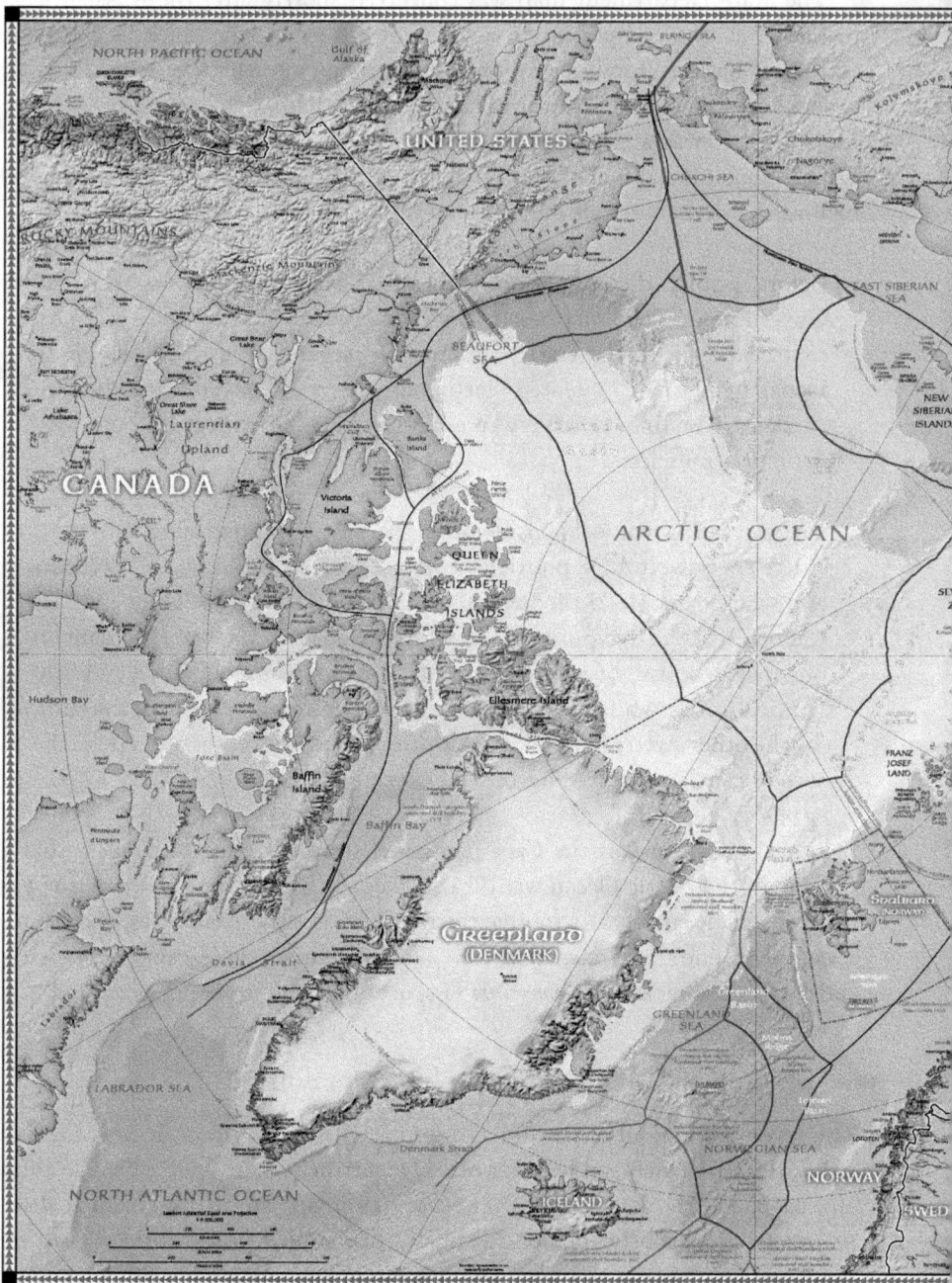

2.5   The Arctic Region, US Department of State and other US Government agencies.

Maritime Lines

Established maritime boundary
Continental shelf boundary
Treaty limit
Claim line
Hypothetical equidistance boundary
Hypothetical passage route
200-NM EEZ (Exclusive Economic Zone) limit

Populated Places

More than 1 million
100,000 to 1,000,000
25,000 to 100,000
Fewer than 25,000
Research station

In this context, it should be recalled that the Russian titanium flag was not the first flag to be planted at the North Pole. Almost a hundred years earlier, in 1909, the American explorer Admiral Robert Peary, together with fellow American Matthew Henson and three Inuit, claimed to have reached the North Pole, and they photographed themselves planting an American flag there. However, if they indeed were at the Pole, they would have been planting the flag not on solid, spatially fixed land but on a mathematically determined spot marked on a maze of mobile and shifting ice floes. When he returned to Newfoundland, Peary sent US president William Taft a seven-word telegram: 'Have honor place north pole your disposal.' President Taft responded: 'Thanks for your interesting and generous offer. I do not know exactly what I could do with it.' It is unclear whether Taft's refusal to accept US sovereignty over the North Pole was due to his recognition that claiming it would have violated prevailing legal norms (e.g., 'I do not know exactly what I could do with it because international law does not permit the United States to claim distant parcels of ocean, even when they are frozen') or if his refusal was based on more practical concerns (e.g., 'I do not know exactly what I could do with it because it is in a remote location with little economic or military value and the costs of maintaining a sovereign presence there would be exceedingly high'). In any event, the flag itself – if still standing – would now be located far from the coordinates that mark the North Pole, carried by the shifting floe ice of the region. The fatuousness of the apparent certainty in this flag-planting is an appropriate symbol of the ambiguous and contested nature of any claim to sovereignty in an icy *terra nullius*.

Natural forces are not the only ones that complicate ideals of claiming Arctic space. According to a Canadian government official whom we interviewed, even if the Russians were on the seabed at the point mathematically marked as the North Pole, they could not have actually placed their flag at that point:

> I heard from a good authority, a Canadian scientist from Victoria, that they did a core of the Pole in the mid-1990s, a soil sample. So he is of the view that [the Russians] didn't actually stick the flag at the Pole because the core of the Pole is in Victoria.

And to add a final complication to the Russian flag-planting narrative, even if the Russian flag actually was planted at the North Pole, it was *almost* not the first flag planted on the seabed there. An individual with a career in both the US Navy and the US Arctic research community related this story in the midst of a discussion about the Russian flag-planting:

I knew there was a great big subsurface buoy at the North Pole installed by the University of Washington. It's placed in 4,300 meters of water, less than 300 yards from the geographic North Pole. And when I heard that this ship was going up there I made an effort through my Navy contacts and the Russian Embassy to try to make sure that this *Mir* submersible knew that the hazard was in fact there . . . .

So we tried to get the word to the Russian government to be careful. You don't want to get that little ship entangled, because it's terribly underpowered and the cable that holds that thing is fairly substantial. It's a Kevlar line; it wouldn't break with the power of a submersible. Well I talked to the scientists about it and it comes within 50 meters of the surface so it's a long string and it's got instruments all over it. And the reason I wanted to stop this thing is because if a submarine ran into it or a surface ship if it was on the surface, a scientist has got so much money tied up in sensors these days that if you lose one of these things, he loses his whole experiment. . . .It's a real hazard, and the scientist loses 6 million dollars. And it's also a hazard to the ship. So it's win win.

So I called Jamie Morrison [at the University of Washington] and I said, 'Well, I tried to get the information on your North Pole environmental observatory to the Russians.' He said, 'Well, I never thought I'd have to tell you this, or tell anybody, but I've got three Arctic bottom pressure recorders at the North Pole that are moored ten meters above the ocean bluff.' And I said, 'Oh shit.' So we went through the drill again. I don't know whether the Russians ever got that information or whether they ever saw these things. Visibility is probably limited to the kind of range you can get with bright, bright light.

But the funny part of the story is that I said, 'Well Jamie, do you have your name and address on those recorders?' He said, 'Yeah we do.' I said, 'Do you have an American flag on them?' He said, 'No, God damn it, but the next one will!'

All three of these stories about flag-planting at the North Pole – the Peary–Taft communication, the University of Victoria coring mission and the University of Washington sensors – as well as the actual case of the Russian flag-planting and its many interpretations, speak to the problems that emerge when one attempts to pair *sovereignty* and *territory* – abstract principles that are based on the control of space – with the actual material nature of the Arctic. The sectoral perspective holds that the Arctic can be claimed, or territorialized, regardless of its nature; indeed, from the sectoral perspective the fundamental divide between land and water that underpins modern notions of sovereignty elsewhere are deemed not to apply to the Arctic. But, as all these stories demonstrate, the nature of the Arctic *does* matter. The Arctic melts and freezes, it moves, and in some cases it is even removed to distant universities for study. In such a mobile environment, the meaning of a planted flag, or a drawn sectoral line, is never beyond contestation. It is unclear whether, prior to the mission, Chilingarov associated a planted flag with a claim to sovereignty. But after the *terra nullius*-fueled media attention that followed the flag-planting, not only the politician Chilingarov, but also the oceanographer Jamie Morrison, understood the various levels of significance that could be accorded to a flag at the North Pole. No wonder, then, that there was such grave concern that, even though it clearly had no legal authority, the flag-planting could still end up meaning *something*.

This chapter has shown the difficulties that emerge when one asserts (or when one implies) that the Arctic is an undifferentiated *terra nullius* that can be claimed by states without regard for its underlying nature. But even when one rejects this perspective, as the international policy community (at least officially) has, it does not follow that the Arctic is necessarily an entirely 'normal' space. Perhaps because the Arctic has a 'different' nature it requires a 'different' form of sovereignty. This is a point that we turn to in the next chapter, and that we will consider again in Chapter 6.

3.1 T-3 (Fletcher's Ice Island).

CHAPTER 3

FROZEN OCEAN

On 16 July 1970, electronics technician Mario Escamilla precipitated an international diplomatic incident when he shot and killed his boss Bennie Lightsy. The facts of the shooting are unremarkable. Escamilla had discovered that his co-worker and neighbor Donald ('Porky') Leavitt had stolen his 15 gallon jug of homemade raisin wine. When Escamilla went to retrieve the jug, he found Leavitt and Lightsy, who was Leavitt's supervisor as well, sharing the raisin wine, blended with 190-proof ethyl alcohol and grape juice. After an unsuccessful attempt to reclaim the jug, Escamilla returned to his residence where, some minutes later, he heard a loud knocking on the door. Because Leavitt was known to be a violent drunk – he had previously come after Escamilla with a butcher's cleaver in pursuit of alcohol – Escamilla grabbed a loaded rifle before answering the door. The man at the door, however, turned out not to be Leavitt but a very inebriated Lightsy. Escamilla and Lightsy resumed their argument and, at some point in the dispute, Escamilla's rifle accidentally discharged, sending a bullet into Lightsy's chest. Despite efforts by Escamilla and others to revive him, within 30 minutes Bennie Lightsy lay dead on Mario Escamilla's floor.

The death of Bennie Lightsy attracted international attention because of its *location*: on T-3, an ice island that from 1952 through 1978 served as a US Navy research station as it floated around the Arctic Ocean. At the time of the shooting, T-3, which was then approximately 21 square miles in area and 100 feet thick, was situated in international waters, about 200 miles north of Canada's Arctic archipelago. Because of its location, it was unclear what state, if any, had jurisdiction to try the case. As a *Time* magazine article reported, Escamilla was caught in 'legal limbo'.

Although a US flag flew over the research station and Escamilla, Lightsy, Leavitt, and the other 16 men stationed there were all American citizens, the United States had long held that the frozen seas of the Arctic were no different than the liquid seas found elsewhere in the world: in all cases they are immune from state appropriation or sovereign jurisdiction, aside from a narrow strip of coastal water where incorporation into state territory is permitted (and even there territorial control must be tempered by allowance of innocent passage by other nations' vessels). This was the same modern construction of the ocean, i.e. as beyond state territory, that subsequently was reiterated in the Ilulissat Declaration, discussed in Chapter 1. The United States was therefore wary of endorsing any solution to the jurisdictional problem that might imply that it had sovereignty over T-3, as this could establish precedent for redefining parts of the ocean as claimable territory. The United States could have argued that T-3, as an ice island consisting of glacial ice that had calved off Canada's Ellesmere Island, was qualitatively different than the much thinner and often seasonal floes of sea-ice that predominate in the Arctic, but even this argument would have had the potential to precipitate an 'ice-grab' that could challenge the 'freedom of navigation' principle that is sacrosanct in US maritime policy.

Canada also faced a dilemma. Canada might have attempted to claim jurisdiction based either on T-3's origin on Ellesmere Island or its location at the time of the shooting in the Canadian sector of Arctic space. However, US–Canadian tensions over Canadian sovereignty in the Arctic were at a high point in 1970, following the 1969 transit of the Northwest Passage by the US icebreaker SS *Manhattan*, which was interpreted by many in Canada as a challenge to its sovereignty. Furthermore, in 1970 Canada passed the Arctic Waters Pollution Prevention Act, under which it claimed the right to regulate Arctic waters well beyond what were then the generally accepted limits of territorial control. In this context, any Canadian claim to jurisdictional authority on T-3 would have further heightened tensions. Canada therefore suggested to the United States that T-3 be considered a vessel, a determination that would have allowed Escamilla to be tried by a US court (because under international law a vessel is considered an extraterritorial extension of its flag state) while avoiding any definitive declaration regarding the territorial status of Arctic ice more generally. The United States, however, had reservations about this proposed solution because it would raise a series of new questions. For instance, if T-3, an 'American vessel', were to become a navigational hindrance and sink a 'real' vessel, would the US be liable for failing to ensure that its 'vessel' was operating in a safe manner?

Trial by the International Court of Justice (ICJ) was proposed by some, but the ICJ is not empowered to try criminal cases involving individuals. Also, at this time, the international law community was preparing to begin negotiations on the United Nations Convention on the Law of the Sea (UNCLOS) and the United States was not inclined to support a position that might set a precedent for placing the ocean (or even a portion of its surface) under the territorial jurisdiction of an international body. US military law was not an option either, because Escamilla was a civilian employee. Another possibility that was briefly mentioned, although apparently not seriously pursued, was for Denmark to claim jurisdiction, because after the shooting Escamilla was first removed by helicopter to Thule Air Force Base in Greenland, before being flown to Dulles Airport in Virginia where he was ultimately arrested.

In the end, the United States claimed jurisdiction over Escamilla not because of where (or what) T-3 was but because of where it was *not*. In its brief to the 2nd District Court in Virginia, the US Government argued that T-3 was effectively a frozen piece of high seas floating on top of liquid high seas and that therefore it was, for legal purposes, beyond territorial control. In such a situation, jurisdiction reverts to the nationality of the individual(s) concerned and since, in this case, both the victim and the alleged perpetrator were US citizens, respectively working for and under contract to the US government, the United States clearly had the most credible national claim to jurisdictional authority. The judge ultimately accepted the case and proceeded with the trial. However, since no formal decision was issued on the matter, it is unknown under what precise grounds the judge accepted jurisdiction: whether it was (1) because all parties were US nationals and the crime occurred in a juridical *vacuum* (a *supra*territorial space); (2) because he classified T-3 as a *vessel* (an *extra*territorial space), or (3) because he classified T-3 as an uninhabited, unclaimed *island*, which would enable the United States to claim provisional possession under the 1856 Guano Islands Act (in which case T-3 was a *proto*territorial space). The judicial panel to which Escamilla appealed following his conviction split 3–3 on whether the district court had erred in hearing the case, and the panel pointedly avoided discussing the topic of jurisdiction in its opinion. Thus, the status of Arctic ice – or at least that of jurisdiction concerning criminal acts on ice islands – remains in 'legal limbo'.

The Legal Status of Ice

Despite the ambiguity that became apparent in the Escamilla case, on a first read the status of sea-ice in international law appears clear: As numerous legal scholars (and several of the individuals interviewed for this book) have asserted, under international law water is water, whether it is in liquid or solid state. As an official at Norway's Ministry of Defence declared, 'The Arctic is an ocean that currently happens to be frozen, but it's water....We should deal with it as if it were a normal ocean,' using UNCLOS as the fundamental framework. Likewise, an American active in the international

Arctic policy-making community stated, 'The Arctic Ocean is an ocean just like any other ocean.' This position is echoed in key Arctic policy documents, such as the Ilulissat Declaration and the 2009 US Arctic Region Policy Presidential Directive.

A United States-based legal scholar, in reiterating this perspective, elaborated on some of the complications that would emerge if one were to ascribe land-like (or, as Canada proposed for T-3, vessel-like) properties to ice:

> Ice, no pun intended, is ... fluid. It melts, so I consider ice [to be] water. I don't consider ice land. The Arctic is largely ice-covered ocean. The North Pole, as a huge chunk of ice, floats on the Arctic Ocean. I don't believe it's anchored to the seabed. It's not a fixed mass. And as we all know it's melting, shrinking in ice. Ice islands are a misnomer; they float. Some people try to say that icebergs could be classified as vessels because they can travel through water. Well perhaps if we had a way to propel them or to navigate with them I could see their point.

The scholar went on to consider the argument, made by some, for using shore-fast ice to define the baselines that are used to determine the extent of territorial waters. Again, however, he demurred from granting this land-like status to ice:

> Ice is not terra firma; ice is not permanent. It breaks off, it calves, and that means the coastline, the ice shelf of the coastline, is going to be constantly changing.... It would be very, I think, impolitic, to set straight baselines using structures that may be gone five years from now.

Others, however, have asserted that the social functions and uses of sea water are so different when it is frozen than when it is liquid that it would be a mistake to simply govern frozen ocean as 'normal ocean', a space that is immune to territorialization. Canadian officials, when arguing for the inclusion of the waters of the Canadian archipelago within that country's internal waters, have been particularly partial to this stance. In 1985, when Canada declared straight baselines around its Arctic archipelago, thereby designating the icy waters between these islands as internal waters, Secretary of State for External Affairs Joe Clark explained to Parliament:

> Canada's sovereignty in the Arctic is indivisible. It embraces land, sea and ice. It extends without interruption to the seaward facing coasts of the Arctic islands. These islands are joined, and not divided, by the waters between them. They are bridged for most of the year by ice. From

time immemorial Canada's Inuit people have used and occupied the ice as they have used and occupied the land.

Twenty-five years later, this justification was reiterated by a senior official with Foreign Affairs and International Trade Canada, who remarked in an interview:

> We're dealing with . . . the world's only large archipelago which has ice-covered areas throughout its surface. The question is what is the status of that ice *vis-à-vis* the land around it, and whether it supports the Canadian view that it is internal waters . . . . At some point we may end up before an international court [and] we will bring evidence that shows the people of the Canadian North – Canadian citizens – in the winter time have treated the ice exactly the same as the land, and we'll make a very strong argument for that.

Several other Canadian officials similarly asserted that the key criterion for assigning sovereignty to space is whether a space is *used*, a concept that was reiterated by Prime Minister Stephen Harper when he proclaimed, in the 2007 Speech from the Throne:

> Canada has a choice when it comes to defending our sovereignty in the Arctic: either we use it or we lose it. And make no mistake, this government intends to use it. Because Canada's Arctic is central to our identity as a northern nation. It is part of our history and it represents the tremendous potential of our future.

For these Canadian officials, the key point is that both land and water in the Canadian North have been used intensively by indigenous Canadians. Therefore, the land–water distinction that is so important elsewhere in the world has less saliency in the Arctic and, it follows, all such spaces that are used by Canadians are integral components of Canadian territory. This special status for Arctic water is seen as flowing directly from its (often) frozen state. As a retired Canadian Forces officer explained, when voicing his support for Prime Minister Harper's agenda and for extending sovereignty over both land and water, 'The Inuit see [ice] as land, and they use it as land.' While anthropologists with whom we later discussed this position decried it as an oversimplification – Inuit perceptions and uses of ice are much more complex than simply seeing and using ice 'as land', and there is even greater variation when one considers the diverse peoples of the North – they agreed more generally with the concept that water, including sea water, in its liquid and especially frozen states, is *internal* to Inuit society in a way that it is not in most Western societies. It follows that, if Inuit are seen as full members of Western nation-states, their uses of the ocean could become the basis for the juridical construction of the Arctic Ocean in a manner different from that which UNCLOS ascribes to the rest of the world ocean.

But for the Arctic Ocean to be classified as an exceptional kind of ocean there would need to be a political will among those who would have the power to transform international law. Instead, most individuals involved in policy debates favor retaining, in the Arctic as elsewhere, UNCLOS' designation of the ocean as a space that is fundamentally different from land and that is (with the exception of coastal waters) beyond territorial appropriation. Indeed, while many individuals interviewed criticized the *process* behind the Ilulissat Conference and the exclusionary way it was convened, few questioned its central message, which was to reaffirm that the Arctic, like the rest of the world, consists of two kinds of surface spaces: land that is apportioned among sovereign states and an ocean that is external to state territory and that is to be governed according to the framework elaborated in UNCLOS.

The (Partial) Exceptionalism of Ice, in Theory and Practice

Notwithstanding the juridical designation of the Arctic as 'normal ocean', many individuals interviewed noted that the Arctic as an ocean nonetheless presents special challenges – for navigation and environmental protection, as well as for legal governance. In most cases, respondents urged that these challenges be addressed through annexes to existing shipping, environmental protection and regional ocean governance agreements rather than through a wholesale reclassification of space in the Arctic's frozen environment.

The model for such annexes is in UNCLOS itself, where Article 234 grants additional rights to coastal states in Arctic waters because of their frozen status. These rights apply not just in adjacent states' 12 nautical miles wide territorial waters but also in their EEZs, which in most cases extend out to 200 nautical miles from the coastline. Article 234 reads, in its entirety:

> Coastal States have the right to adopt and enforce non-discriminatory laws and regulations for the prevention, reduction and control of marine pollution from vessels in ice-covered areas within the limits of the exclusive economic zone, where particularly severe climatic conditions and the presence of ice covering such areas for most of the year create obstructions or exceptional hazards to navigation, and pollution of the marine environment could cause major harm to or irreversible disturbance of the ecological balance. Such laws and regulations shall

have due regard to navigation and the protection and preservation of the marine environment based on the best available scientific evidence.

Article 234 avoids any references to *uses* of ice and thus has little in common with the previous statements from Canadian officials who suggested that the land-like properties of sea-ice challenge the fundamental division of space between land and water. Rather than addressing indigenous peoples' uses of sea-ice, Article 234, as Canadian anthropologist Claudio Aporta has noted, is written from a perspective informed solely by Western navigation: ice is viewed as a 'hazard' and an 'obstruction', a negative property that is associated with 'severe climatic conditions'. Sea-ice is affirmed as existing beyond state territory, the same as liquid ocean. However, because it is recognized as being more dangerous than liquid ocean it is deemed suitable for an extraordinary level of state governance.

While the implications of Article 234 for state sovereignty and ocean governance appear straightforward, in fact numerous controversies have surrounded its application. Article 234 is the only article (out of 320 articles) in UNCLOS that specifically mentions ice, and there is nowhere else in the convention that one can go to for guidance regarding what precisely is meant by 'severe climatic conditions' or 'the presence of ice cover ... for most of the year'. This issue was specifically addressed in an interview with a former employee of Canada's Ice Service, the office that assesses ice cover and advises mariners and the Canadian Coast Guard on conditions along Arctic shipping routes. The Ice Service's standard for identifying a region as being 'ice covered' for 'most of the year' is that it has at least 10 percent ice coverage for at least six months of the year. However, during the interview it was revealed that the association of 10 percent iciness with the term 'ice covered' was not universally accepted by all Canadian governmental offices. The former Ice Service official remarked,

It's interesting, because twenty years ago there was a discussion of the use of the term 'ice covered.' And the decision was that we really shouldn't use the term 'ice covered' to describe Canadian waters; that's seen as a negative.

Other terms, such as 'ice encumbered' and 'ice infested' were proposed, so as to better communicate that in many cases these icy waters were still primarily liquid and therefore potentially navigable. More recently, he went on to note, Foreign Affairs and International Trade Canada had recognized the value in classifying large portions of Canadian waters as 'ice covered' so as to permit application of Article 234 and Canada's companion legislation, the Arctic

Waters Pollution Prevention Act (AWPPA). Additionally, although it is not specifically countenanced by Article 234, there was a general belief among individuals at Foreign Affairs that emphasizing the icy state of the region's waters could enhance Canada's claims to declaring the waters of the Canadian archipelago as 'internal waters', as illustrated by the statements presented earlier in this chapter from Secretary of State for External Affairs Clark and the Foreign Affairs official. For these reasons, according to the official, 'We are kind of retreating back to saying, "They are definitely ice covered most of the year".' He concluded by noting that the ebbs and flows of this debate over the terminology and threshold for defining ice-covered waters shed some light on 'where the pressures have changed over the years'.

This potential for extending claims to enhanced governance based on Article 234 into something resembling an assertion of territorial sovereignty over 'land-like' frozen waters – whether individual ice islands like T-3 or Arctic sea-ice more generally – has been especially troubling to the United States. Of all the Arctic nations, the United States has been exceptionally vigilant in guarding against what it perceives could be threats to the freedom of navigation that plays a key role in the United States' projection of power around the world. Turning specifically to the Arctic (and alluding obliquely to Canada in particular), Commander James Kraska of the US Navy has written,

> Some governments have taken the view that the ice can be occupied, converting frozen water into a sort of 'ice territory' with attendant rights. This is a purely theoretical invention and has no basis in either customary international law or the Law of the Sea Convention.

Indeed, the United States has a long history of protesting against attempts to territorialize Arctic waters, from its disavowal of Admiral Peary's offer to claim the North Pole for the United States in 1909 to the protests that it has filed in recent years against each of the other four Arctic coastal states' straight baseline and historic waters claims. In 2010, the US State Department filed a diplomatic note objecting to Canada's mandate that other nations' vessels participate in NORDREGs, a Transport Canada-maintained registry for vessels entering Arctic waters in the vicinity of Canada (including waters of the EEZ, where coastal states normally have no authority over navigation). The diplomatic note asserts that the United States finds making NORDREGs participation mandatory to be an excessive interpretation of Article 234. In its protest, the United States restates its position that similar protections for environmentally sensitive Arctic waters could be achieved through strengthening the International Maritime Organization's code for ships operating in polar waters, a regulatory mechanism that would be just as effective and that would avoid any implication that Arctic waters are susceptible to territorial appropriation by states. The diplomatic note concludes with the following paragraph, which reveals the depth of the United States' objection:

> The United States noted with concern the references to 'sovereignty' in the statements accompanying the regulations. The United States wishes to note that the NORDREGs do not, and cannot as a matter of law, increase the 'sovereignty' of Canada over any territory or marine area.

In fact, both Canada and Russia have national legislation that specifically addresses the icy nature of Arctic waters. In both cases, although the stipulated reasons for this legislation — as with Article 234 — are the environmental challenges posed by ice-covered waters, their history is interlaced with national ideals, discussed in Chapter 2, regarding the incorporation of sea-ice within sovereign state territory and beliefs about the inapplicability of temperate-zone norms that divide land from water in the frozen North. In Canada, the AWPPA permits Canada greater leverage to enact environmental controls in an exceptionally vulnerable environment, originally up to 100 nautical miles from its coast, a limit that in 1985 was extended to 200 nautical miles so as to align with the extent of Canada's EEZ. However, the AWPPA was promulgated in direct response to the politically contested passage of the US icebreaker SS *Manhattan* through archipelagic waters and its passage in 1970 also served as a means by which Canada could assert control over an Arctic maritime sphere of influence without directly challenging the position held by the United States (and others) that Arctic waters are fundamentally like any other waters. The references to 'sovereignty' in the justifications given by Canada for declaration of mandatory NORDREGs notification, as well as the United States' reaction, imply that Canada's attempts to assert sovereign power in adjacent waters beyond the 12 nautical mile limit of its territorial sea continue to be recognized by all as occurring within a highly politicized debate regarding the application of temperate-zone norms of ocean governance to a primarily ice-covered environment.

The AWPPA was passed in 1970, three years before UNCLOS negotiations formally began and, when negotiators started to meet, Canada took a leading role in arguing for an AWPPA-like provision in UNCLOS. This ultimately became Article 234, a provision that grants special — albeit not sovereign — rights to protect the environment of ice-covered waters within their EEZs. Canada's major ally in this effort was the Soviet Union.

Although the Soviet Union (and, later, Russia) did not enact legislation with
the scope of the AWPPA – comprehensively differentiating Arctic from
non-Arctic waters and bestowing the state with special powers in these
Arctic waters – the USSR/Russia, too, has seen a number of Article
234-inspired Arctic-specific maritime regulations since the finalization of
UNCLOS in 1982. These include the 1984 Edict on Intensifying Nature
Protection in Areas of the Far North and Marine Areas Adjacent to the
Northern Coast of the USSR and the 2012 Law on the Northern Sea Route.

In contrast, US ocean law makes no distinction between Arctic (or frozen)
seas and other waters.[1] When asked why this was so, American respondents
gave a range of interpretations. An official with the US Environmental
Protection Agency contrasted the US regulatory environment with that
encountered by his counterparts at Environment Canada. In the United
States, he remarked, there was no need for a special law for Arctic waters
because statutes were written with enough flexibility to allow for place-
specific permitting rules that reflected local conditions:

> Whatever statute drives [a permitting process] will accommodate
> dealing with local conditions. So, to the degree that you have concerns
> about ice floes or ice movements or depth, or whatever you're dealing
> with [in the Arctic], our permits are flexible enough .... [For instance,
> in permitting oil drilling infrastructure,] we use the same set of rules
> down south in the Gulf [of Mexico] as [the ones] we use in Alaska. It's
> just that they are adapted for different [regional conditions.]

Others, however, perceived this difference between the American and Canadian
approaches to Arctic waters as emanating not from different domestic legal
environments but from differences in the two countries' positions in the
international arena. An officer with the US Coast Guard speculated that the
failure of the United States to address the specific circumstances of Arctic waters
through targeted frozen-waters legislation is due to its failure to accede to
UNCLOS. While the United States follows UNCLOS as customary law, he felt
that any frozen-water-specific legislation would so clearly be derived from
Article 234 that 'we would get beat up [by] the international community [who
would ask] "How can you do that when you aren't even a state party to the
Convention?"' A retired US Coast Guard officer added another layer of
complexity to this argument. He pointed out that coastal states (like Canada)
that have ratified UNCLOS can assert increased authority in adjacent frozen
waters and point to Article 234 to clarify how this higher level of authority does
not constitute a claim to these waters as sovereign territory. If, by contrast, the
United States were to adopt a law like the AWPPA without first acceding to

UNCLOS, then it would be difficult for the United States to communicate to the world community where this falls short of territorialization, the division of a space into bounded and governed territories controlled by individual states. As a result, if the United States were to enact frozen-water specific ocean governance legislation it could inadvertently precipitate the territorialization of sea-ice that, as was demonstrated in its position in the T-3 case, is inimical to US maritime policy. This individual went on to speculate that *if* the United States were ever to accede to UNCLOS it might then follow up with AWPPA-like legislation to govern its adjacent frozen waters.

In fact, these concerns that Article 234 could lead to a kind of creeping territorialization of the Arctic Ocean are present even in states that have enacted frozen-ocean specific legislation. This point was made particularly clearly by an official at Fisheries and Oceans Canada, who remarked that even though Canada had ratified UNCLOS and adopted its own Arctic-specific legislation in concordance with Article 234, she was concerned about the openings created by highlighting the uniqueness of Arctic space:

Respondent: There are some principles we would stand strongly by. One is international law ... the really foundational law. We will never agree to tinker with that because we have more to lose than gain by tinkering with it to accommodate Arcticness. So we would always deal with [the specific needs of Arctic space] through an annex or guidelines or something like that .... We have taken the lightest hand possible that we can in terms of global concessions because we're never, ever going to agree in my lifetime [to a comprehensive international legal regime for the Arctic]. I can't see us ever going to do anything there, because the risks on so many other agendas are too great.

Interviewer: It has a history with Article 234 though.

Respondent: I know, but there's too much baggage. There are too many other agendas at play that are just looking for an opening .... Things are packaged, and you can't control the package. So once you go down that road, you are opening everything up. So we will be dealing with guidelines, we'll be dealing through implementing legislation, stuff like that, but we wouldn't be tinkering with the law itself.

Security and Development on Northern Frontiers

Canadians may be especially attuned to both the needs for and implications of highlighting the specificity of the Arctic environment because Canada's northern policy specifically recognizes the complementarity of Arctic security with the goal of cultural and economic development on the nation's northern frontier. Canada's Northern Strategy, whose maps were discussed in Chapter 2, was released in 2007 and combines one outward-looking dimension – 'Exercising our Arctic sovereignty' – with three inward-looking dimensions – 'Protecting our environmental heritage', 'Promoting social and economic development', and 'Improving and devolving northern governance'. Indeed, a common theme raised by Inuit leaders, as well as members of the opposition Labour and New Democratic Parties, when critiquing the military focus of the ruling Conservative Party's northern policy, is that the best way to buttress Canadian sovereignty in the North is through the economic development of Canadian northern peoples and territories. In an interview, one Inuit elected official and Labour Party activist directly co-opted Prime Minister Harper's 'use it or lose it' line, pointing out that any policy based on asserting northern sovereignty through 'use' must involve not just securing Canada's (land, water and ice) borders but also developing its (land, water and ice) territories and the people who inhabit (and thereby 'use') them.

In Russia, this pairing of economic development with national security has historically had a somewhat different tone, since economic development there is generally associated more with settlement by southern migrants rather than incorporation of indigenous northern peoples. Nonetheless, as in Canada, the military and economic agendas have been linked in Russia in an effort to 'secure' the northern frontier, especially since 2007 when northern issues began to take on a new centrality in Russian nation-building rhetoric. This can be seen in efforts such as the 2011 establishment of the Northern (Arctic) Federal University in Archangelsk, whose mission joins a regional development agenda with one that reaffirms Russia's status and security as an Arctic nation.

In the United States, by contrast, there are two largely autonomous sets of policies directed at the North: 'Arctic policy', which is focused on the maritime arena of international affairs and is devoted to reproducing global principles of maritime freedom; and 'Alaskan development', which is focused on land and nearshore areas and which is, depending on one's persuasion, devoted either to promoting oil and gas development or preserving the state's wild and pristine nature. The disconnect between 'Arctic' and 'Alaskan' policies in the United States was elaborated on in an interview with an Alaskan who had attended several meetings of the Arctic Council and other international pan-Arctic fora:

> At my very first meeting of the Arctic Council, I was shocked that there were no Alaskans. And there have been many other meetings where I was the only Alaskan at the table ….[It is important] to bring

Alaskans, I'm not even talking about indigenous, but Alaskan interests [to these meetings]. Because, I mean it's not a secret that people from the State Department, they don't know much about Alaska. And Alaska is, it's an overseas territory . . .

Denmark [always] has a Greenlandic representative and a Faroe Islands representative at the table, and they have three flags. And I've been trying to see if there is support for having an Alaskan flag next to the US, which I think would be appropriate. Alaska is a special state . . . But the State Department is very sensitive. For example, on [circumpolar] maps, [the United States] is often indicated just by Alaska. You know, you have all of the countries and then Alaska. And every map from the State Department will cross out 'Alaska' and put 'US' . . . I do not think that there would be a real danger if Alaska was given a little bit of its own place.

Indeed, in the 2009 US Arctic Region Policy Presidential Directive – a document that was drafted by an interagency taskforce but under the leadership of the State Department – the word 'Alaska' never appears. Although the document asserts, 'The United States is an Arctic nation', several respondents in Alaska noted that from their perspective this acknowledgment is not borne out in the attitudes or policies of Washington-based officials.

Respondents in Norway were even more explicit about the separation between Arctic strategy and northern development. The following transpired during an interview with two officials at the Ministry of Defence:

Respondent 1: I think a certain challenge for the government is that people up in the High North feel that . . . the current [High North] strategy would change something for them, too, because there's a certain expectation that something would happen, for example in [schools], or universities or whatever.

Respondent 2: That's absolutely right. I think there are two possible approaches internally in Norway. One is that the Arctic or High North policy is a part of everybody's and is a part of Norway's policy, which is a broad perspective. And the more narrow perspective is that it's another name for an enhanced, discrete development policy. Of course they are related but I think sometimes we have felt, and the foreign minister has said this clearly, that sometimes he goes to the North and he's met with a comment, 'We thought we'd get new roads because of the Arctic

policy.' Well, I wish you luck with having roads but that's not exactly
what the Arctic policy is about. It's more about the strategic interest.

In a similar vein, when a member of the Norwegian parliament was asked how
the presence of Saami in the High North complicated the development of a
Norwegian Arctic policy, the parliamentarian responded that their presence
forced the government to engage in 'discussions' regarding distribution of
resource revenues and that the Norwegian government encouraged the Saami
to take a leading role in circumpolar indigenous peoples' forums. But the
respondent failed even to consider how the presence of indigenous peoples
might impact Norway's Arctic policy since, for him, like the Defence
Ministry respondents, Arctic policy revolves around 'the strategic interest',
rather than the apparently secondary goals of economic or social development.

Merging National Legal Systems with Indigenous Uses

These different national perspectives on the relationship between the Arctic as
a space of national development and the Arctic as an international maritime
arena in turn are reflected in the degrees to which nations are willing to adjust
their national laws in order to accommodate the different ways in which
indigenous peoples integrate water (including sea-ice) into their lifeworlds. In
the United States, the federal government's primary concern has been to avoid
making any accommodations that might challenge the fundamental division
between land and water that is the basis for the doctrine of freedom of
navigation. This was seen in the United States' approach to the T-3 incident
but it also can be seen in the federal government's policy on indigenous
peoples' issues. For instance, our interviews with Alaskan Aleut activists
revealed a high level of frustration with the US government's failure to grant
the Aleutian islanders territorial control over the seas among their islands,
even though, according to the Aleuts, the 1867 Treaty of Cession between the
United States and Russia gave indigenous peoples control of sea territory as
well as land.[2]

In Canada, by contrast, there is a strong acknowledgment that the Inuit are
a maritime people and that this must be reflected in indigenous self-
governance regimes. Referencing the fact that all but one village in the
majority-Inuit territory of Nunavut is on the coast and that Inuit livelihoods
are as dependent on water and sea-ice as on land, an official with a Canadian
Inuit rights group went so far as to proclaim, 'People aren't interested in the
land, it's obviously just the water.' A non-Inuit politician who worked for the
federal government in negotiating the Land Claims Agreement that
established Nunavut explained how, during the negotiations, explicit links

3.2   Focus Group in Pangnirtung, Nunavut, Canada (sea-ice map and terminology verification).

were made between improving the livelihood and security of Inuit peoples and reaffirming (or, perhaps, extending) the sovereignty of the Canadian state over land *and* sea:

> [When negotiating,] we felt very strongly that the marine economy and tradition of the Inuit had to be reflected in the land claims. And it is writ large in the land claims, there are sections of the Land Claims Agreement that credit the Inuit with having assisted in establishing Canadian sovereignty; [that is] explicit in the preamble of the Inuit land claims. The Inuit carved out a jurisdiction based on the fast ice, an offshore jurisdiction based on the fast ice that kind of clings to the land. And also the land claims called for a Marine Council that sadly has never been implemented.

Recognizing that they were expanding the meaning of state territory in new directions that might ultimately conflict with Foreign Affairs' position that officially reproduces international norms about the external nature of ocean space, negotiations on the portions of the Nunavut Land Claims Agreement that established the basis for Inuit sea rights proceeded with difficulty:

The federal government, in the negotiations, was very cautious and
guarded about doing anything, about conceding anything but the most
general interests in the offshore. There was a sense that this was a land
and mining claim and land-based, and so the Land Claims Agreement is
not detailed or specific about an Inuit role in offshore management,
except in provisions like the Marine Council.

The politician went on to note that the ambiguous nature of what it means
to have a sea claim – and the vague way in which it is taken up in the Land
Claims Agreement – has in turn resulted in the current stalemate, under
which the Nunavut Marine Council has failed to be established. On one side,
he noted, is Nunavut Tunngavik Incorporated, the successor to the Inuit
organization that negotiated the Land Claims Agreement (Tunngavik
Federation of Nunavut), which maintains that Inuit should have special rights
to marine resources because of the ways in which they have traditionally
integrated the sea into their lives. The Government of Nunavut, on the other
hand, bases its arguments for marine management rights on precedent set by
southern provinces that have established co-management boards and revenue-
sharing arrangements, a position that effectively reproduces notions of the
Arctic as a 'normal' space with a 'normal' separation between land and sea.
This debate, in turn, reflects an internal conflict within Canada between
those who, like the Foreign Affairs and Canadian Forces respondents quoted
earlier in this chapter, highlight the unique integration of sea with land in
Canadian territory so as to buttress claims to Canadian sovereignty in the
Arctic, and those who, like the Fisheries and Oceans respondent quoted
earlier in this chapter, fear the 'agendas' that could be unleashed in the
international arena if one were to adjust normative laws to account for the
specificity of 'Arcticness'.

Two aspects of this debate within Canada bear mentioning. The first is
that it is happening at all. This contrasts markedly with, in particular, the
United States, where – in part because of US concerns with setting a new
precedent for other regional seas and in part because of the marginal role of
the Arctic in national consciousness – any accommodation of established
norms to the material conditions of the Arctic or the material realities of
Arctic livelihoods is off the agenda. The second point is that material
conditions in the Arctic *do* present unique challenges for a range of users.
Just as industries have found a need to adapt temperate-zone derived
technologies to the harsh realities of the Arctic environment, perhaps
similar adaptations are required for temperate-zone derived laws. Thus, a
strong argument can be made for *some* degree of exceptionalism, even if
only at the level of interpretation or implementation.

Governing Ice amidst Climate Change

For many users of the Arctic, the unique challenges of the Arctic environment will only increase with climate change. With reference to shipping, one expert noted,

> Climate change, which is translated for most of us in the marine world as sea extent and thickness, is not really the driving factor [in the increased attention being directed at the Arctic]. It's nice that ice is...thinner and extent is going away for a couple months of the year when it's ice free, but for most of the year, for most of this century and beyond, the Arctic Ocean is [going to be] ice covered. The ice [will be] thinner, but I think the notion that it's [going to be] an ice-free Arctic Ocean [is wrong] .... This has implications for the regulation of ships, they'll have to be polar class ships, they'll be more expensive. And then of course ship speeds [will be affected]. If the ice is thick enough it will negate the time–distance saving.

In some parts of the Arctic, including the Canadian archipelago and the east coast of Greenland, melting ice and decreased ice thickness could actually create *more* hazards for shippers and offshore oil and gas extractors, as ice is freed from its moorings, increasing the numbers of ice floes and icebergs that clutter emerging waterways and endanger drilling platforms.

Although respondents in the oil and gas industry stressed that the likelihood of a blow-out on the scale of the 2010 Deepwater Horizon incident in the Gulf of Mexico is reduced because the Arctic's relatively shallow waters would facilitate drilling relief wells, seasonal conditions in the Arctic – which will continue for the foreseeable future – would complicate the response to any major event that were to occur. As one Alaska-based industry regulator remarked with reference to offshore drilling:

> A blow-out in the Arctic, if it were to last into the freeze up, would be extremely challenging and potentially very damaging .... There is [only] a tiny little window [for drilling and for taking remedial measures in the summer season]; they have to be off the well long before freezing.

A US Coast Guard official added, 'There's not a whole lot of good methods for cleaning up oil on [sea] ice.'

Onshore, the situation is somewhat different. Whereas offshore drillers view the coming of frigid weather with apprehension, onshore drillers find that their risks increase when the ice begins to melt. As a veteran of the oil spill remediation industry noted:

Spills on the ground in the winter time are the easiest and the best to clean up. If you're going to have a spill, you might as well have it when everything is still frozen. It won't penetrate the tundra. It really has to be a huge spill to get to the point [where it causes major problems].

In addition, melting permafrost destabilizes equipment, increased temperatures reduce the season for ice roads and the decline of sea-ice leads to the disappearance of erosion-preventing ice barriers.

Beyond the oil and gas industries, climate change is causing a host of difficulties for life in Arctic regions. A government official from Canada's Northwest Territories enumerated these:

I remember I made a presentation for an environment committee in Ottawa 20 years ago and ... the first question they asked me [was], 'Is it true that you people in the North can't wait until global warming occurs,' and my answer was, 'No, it is not a good thing.' It is changing our environment, it is changing our way of life, it is negatively impacting almost everything, transportation systems, rivers, water, the ice is melting, ice seasons are changing, we are losing all kinds of species, species that have never been north are moving north, and we're starting to see weird things like hybrid bears. The water – species of fish are disappearing from the Mackenzie River because the water is getting too warm .... That is what we are looking at, our expectation is that we are going to have very drastic forest fire seasons as the tree lines are moving further north, and we are already seeing that there is a continuous and discontinuous permafrost. The continuous permafrost may very well be disappearing, the permanent permafrost is starting to melt as well, so it is affecting our buildings. Historically we used wood pilings, we would put the wood pilings in the permafrost. All the wood pilings are rotting, so we have to go to stainless steel, which is very expensive.

Clearly, the challenges of northern climates, all of which can be used to support the argument that the region is deserving of special protections and legal regimes that push against the limits of the land–sea divide that characterize the rest of the world, persist even under climate change.

The Challenge of Remoteness

Notwithstanding these issues caused by the Arctic's climate, arguably the most compelling reasons for modifying legal norms to accommodate the specific characteristics of the Arctic environment rely less on the region's iciness than its remoteness from southern centers of population. Several respondents referenced remoteness during the interviews. However, the logistical difficulties caused by remoteness, distance and poor infrastructure – in contrast with the relatively manageable problems of ice – were highlighted in particular by the director of an Alaska-based oil spill cleanup concern. This individual was interviewed in 2010, when many of his employees were working in Louisiana on the Deepwater Horizon effort:

> It is a logistical nightmare down there [in Louisiana]. And they have roads, they have hotels and restaurants where you can eat. Take that same scenario and bring it up to the North Slope where you have got a finite area for your infrastructure of roads and access to the water, and then you find out how you're going to support something that is potentially across a wide swath of places where there are no roads, there is no place to drive to, you have to get there by boat or fly in. Do you have enough of those things to get you where you need to go? So a lot of logistical support kinds of things are probably going to come up ... I don't think anybody could really respond to something that big up here in this remote area, it would be tough ....

> Down in the Gulf [of Mexico], you've got hazards: it's hot down there. So our guys are working for 15 minutes and then cooling down. Up here in the wintertime we work 15 minutes and we have to warm up. We're used to that kind of stuff. But as far as the remoteness of getting around and being able to support [ourselves], that [is where new issues arise.] We have polar bear issues; you're not at the top of the food chain up here. So if you're putting folks in remote areas you've also got to think about personal protection, aside from survival. We are confident that we can do our job up here in the wintertime, but if you [have a spill on the order of] something like 10,000 barrels a day or whatever is going on [with Deepwater Horizon], I don't think anybody could be ready for something like that [up here].

It follows that if the Arctic needs to be governed by a region-specific regime, the key factor behind its exceptionalism may not be the presence of ice but rather the region's extreme remoteness from state-based institutions of civil authority, including, but not limited to, logistical infrastructure.

Indeed, to return to the story with which we began this chapter, it was the *remoteness* of T-3 – not its geophysical existence as a mobile chunk of glacial ice – that led the United States Court of Appeals to overturn Mario Escamilla's conviction for involuntary manslaughter. In its decision, the

Court of Appeals criticized the judge in the original trial for highlighting the *unexceptional* nature of T-3. The trial judge in the District Court had instructed the jury:

> The law of the United States is applicable to this case in identically the same manner as it would be applicable to the same crime if it were committed right here in Northern Virginia. So you can just forget the [T-3] part other than for background.

This, the Appellate Court declared, was an erroneous instruction:

> It would seem plain that what is negligent or grossly negligent conduct in the Eastern District of Virginia may not be negligent or grossly negligent on T-3 when it is remembered that T-3 has no governing authority, no police force, is relatively inaccessible from the rest of the world, lacks medical facilities and the dwellings thereon lack locks – in short, that absent self-restraint on the part of those stationed on T-3 and effectiveness of the group leader, T-3 is a place where no recognized means of law enforcement exist and each man must look to himself for the immediate enforcement of his rights. Certainly, all of these factors are ones which should be considered by a jury given the problem of determining whether the defendant was grossly negligent.

The exceptionalism of the Arctic, then, may lay not only in its frigid weather, or its geophysical specificity that challenges the norms of international law, as in what, from the perspective of southern capitals, is its inaccessibility. As a space that will always remain distant (and different), the Arctic is arguably not a *frontier* but a *colony*: not a space to which distant states and their populations expand but a space that they colonize so as to benefit the metropole. In such a situation, the appropriate political response might focus less on adjusting existing laws to grant distant capitals special powers over their icy domains and more on granting these domains political independence. It is to this imaginary – and in particular that of Inuit statehood – that we turn in Chapter 4.

4.1 Greenland National Day, Nuuk, Greenland.

CHAPTER 4

INDIGENOUS STATEHOOD

In a 2007 cable from the US Embassy in Copenhagen that was released via Wikileaks it was opined,

> Greenland is on a clear track toward independence, which could come more quickly than most outside the Kingdom of Denmark realize . . . . One senior Greenlandic official commented recently that his country (Greenlanders and many Danes alike routinely refer to Greenland as a 'country') is 'just one big oil strike away' from economic and political independence.

Greenland has steadily been moving towards greater autonomy from Denmark, and with the prospect of potentially massive oil reserves off its coasts there suddenly appears to be a viable financial option to make complete independence possible. Such a turn of events would appear to have transformative potential, involving the actual redrawing of boundaries and the creation of a model for indigenous statehood that could be replicated elsewhere in the Arctic (most notably in Canada's Inuit-majority territory of Nunavut).

Yet, an independent Greenland would face dilemmas similar to those faced by other small, postcolonial dependencies: How can formal political independence be translated into economic independence, especially in a region like Greenland where the main potential source of economic wealth – offshore oil and gas extraction – would be capital intensive and heavily dependent on outside investment and technology? Indeed, it is telling that the above-quoted US State Department cable goes on to herald Greenlandic independence as a 'unique opportunity' for US commercial interests. As we look to the broad contours of governance in the Arctic region, the long-term impacts of Greenlandic independence may appear less significant: Greenlandic independence would reproduce the basic structures of the state system, including the status quo outlined in Chapter 1, and the 'race' for resources would in no way be hindered. In fact, it is likely that the needs of a new independent Greenland would call for a ratcheting up of resource exploitation.

Nonetheless, a number of key questions emerges: Would Denmark permit Greenland to secede after it finds a treasure trove of hydrocarbons off its shores? What would Denmark's role in the Arctic be if Greenland became a sovereign state? Would the Arctic region around Greenland become the site of massive oil and gas infrastructure with immense strategic value? What geopolitical implications would set in, given such economic developments? And lastly, what would these developments mean for the broader aspirations of native peoples across the circumpolar region? Is Greenland, which is majority Inuit, truly a model for other indigenous groups in the Arctic? All of these uncertainties are still hanging in the air, as Greenlanders hold their breath in anticipation of that big oil strike.

A Brief History

Although the first paleo-eskimo peoples settled Greenland as far back as 2500 BC, it was not until roughly 1300 AD that the Inuit Thule people, their origins reaching back to Alaska, migrated into northwestern Greenland. At that time the paleo-eskimo Dorset culture still inhabited this part of Greenland, and Norse (Norwegian/Icelandic) settlements, which had been established 300 years earlier, were located in the south. Eventually the Dorset culture was supplanted by the Thule, and in the fifteenth century the Norse settlements in the south proved to be unsustainable. Norway, however, not knowing what had happened to the Greenlandic colonists, maintained its claim on Greenland. It was Denmark, after joining a union with Norway, that sent Hans Egede to Greenland in 1721 to uncover the fate of the earlier colonists and to establish a commercial and religious mission there. His mission, established at the tip of a fjord in southwestern Greenland, became the settlement of Godthåb ('Good Hope'), which eventually would become Nuuk, the current capital of Greenland. From this outpost the Danish–Norwegian Kingdom was able to expand its influence and establish more settlements along the western coast. In 1774 the Royal Greenland Trading Department was established, maintaining a monopoly over Greenlandic trade for nearly 200 years, focusing particularly on trade with Inuit for whales and seals.

The first break with Danish rule on the island came during World War II, when the United States occupied Greenland after Germany's invasion and occupation of Denmark. During this time, Greenlanders, whose contact with the outside world had been strictly controlled by Denmark, began to develop initial self-government institutions and a growing sense of autonomy.

The United States, in turn, had become interested in the geopolitical value of Greenland and offered to buy it for $100m in 1946. Denmark refused this offer and it responded to the internal developments within Greenland by establishing it as an integral part of the Danish Kingdom in 1953. It should be noted, however, that for Denmark Greenland was never a money-making territory, with most economic activity limited to fishing and shrimping.

Like most other governments with Inuit populations at this time, Denmark attempted to assimilate the Inuit people and create Greenland as a replica of itself. The welfare state principles of Denmark were applied to Greenland and the population was centralized into larger settlements where government services, including education, could be delivered. This marked a fundamental change in the lives of the Greenlandic Inuit who, due to removal from their subsistence lifestyles, were becoming dependent on urban-based activities, especially the government sector. Attempts were made to impose the Danish language on all Greenlanders and upcoming Inuit elites were encouraged to obtain university education in Denmark.

For their part, however, Greenlandic Inuit were becoming increasingly aware of, and resistant to, the loss of their own culture. Discussions on autonomy – and even independence – in Greenland can be traced back to the mid 1800s but, with Danish assimilation policies taking full effect in the 1950s and with the movement to exempt Greenland from the European Economic Community (EEC) in the 1970s, Greenlandic national conscious-ness was being raised like never before. The educated Inuit elites returning from Copenhagen, while having absorbed significant aspects of Danish culture and values, did not consider themselves purely Danish and demanded a greater say in Greenlandic affairs. In response to this increased pressure, Denmark granted Greenland home rule status in 1979, which, for the time, provided significant autonomy to the Greenlanders. A local parliament empowered to decide over most internal affairs was established and Greenland was allowed to exit the EEC in 1985.

By this time, demands for Inuit empowerment and calls for recognition of a circumpolar Inuit nation were growing across the Arctic, as we discuss in detail in Chapter 6. In this context, Greenland's attainment of home rule was seen as an immensely progressive step for the circumpolar Inuit movement. Yet demands grew, and in 2008 Greenlanders voted in a referendum to expand even further the autonomy laid out in the home rule regime. The subsequent Act on Greenland Self-Government essentially achieves the aims of self-determination as outlined by Inuit empowerment organizations across the Arctic, such as the Inuit Circumpolar Council (ICC), although, it should be added, the decree, like that which ten years earlier had established the Canadian territory of Nunavut, entails no explicit mention of the ethnic or indigenous character of the Inuit. The Act includes (1) the recognition of Greenlanders (defined as anyone born in Greenland) as a distinct people under international law; (2) the further empowerment of a Greenlandic government, both executive and legislative, with expanded jurisdiction over the Greenlandic police and courts; (3) full rights over the soil and subsoil; and (4) expanded power to be engaged in foreign affairs that may have repercussions for Greenland. Of all of these points, clearly the biggest coup in the self-rule agreement was the attainment of full rights over the soil and subsoil, since this gives the Greenlanders full control over any potential oil and gas finds on the island's territory or in its adjacent waters.

The move to self-government, which was widely supported in Greenland, has contributed to a confident outlook among the Greenlandic leadership. In interviews conducted with Greenlandic officials there pervaded a sense that Greenland is semi-sovereign, as opposed to the earlier arrangement whereby Copenhagen was still seen as the political center pulling the strings. As a Greenland political representative in Copenhagen put it,

> Earlier, somebody could have [said], 'Ok, if things don't work out, Copenhagen will take over.' But that is, from a political and legal point not the case anymore. . . And that, as I see it, has brought kind of a new energy, a new sort of spirit into the political system. That does not mean that our problems are gone. But it has been very important developmentally speaking.

Yet, despite these achievements, which have been much admired by fellow Inuit in the United States, Canada and Russia, the predominately Inuit leadership in Greenland has now fixed its gaze on a further, though still distant, goal: full independence. As the same official quoted above confided: 'At the moment, getting independence, at any price, is at the top of the political agenda.' Or, as another Greenlandic official remarked, 'There is the goal where we all want to become independent. Throughout the whole political spectrum, you don't talk about Greenlandic independence like something that might happen, or if it happens. It is a question of when'.

Independence

There are a number of dimensions to Greenland's independence ambitions that are relevant to the governance future of the Arctic. To begin, Greenland's desire for independence is firmly rooted in an acceptance of the modern state system that was outlined in Chapter 1: that is, the political organization of the world into discrete, territorially defined, sovereign entities called nation-states. Although this system is largely taken for granted by everyday citizens, the preeminence of the nation-state has nonetheless come under duress in the wake of globalization and the emergence of powerful transnational corporations and social movements. One of the most relevant social movements in the Arctic, and one with particular importance for Greenland as a majority-Inuit territory, has been that of Inuit self-determination. As we discuss in Chapter 6, the transnational Inuit movement has refused to fully accept the dictates of the state system. The Inuit Circumpolar Council (ICC) has been the leading transnational organization to represent the Inuit in this movement, although there are many other non-governmental organizations (NGOs) and indigenous corporations, particularly in the United States and Canada, working on local and regional levels that have similarly embraced the cause of an Inuit right to self-determination that transcends the modern nation-state.

Greenland, however, whose population is more than 85 percent Inuit, is finding itself increasingly at odds with the governance framework espoused by the ICC due to its embrace of the nation-state ideal. In this sense Greenland sticks out, as no other Inuit territories have so directly pursued the status of statehood. In Alaska, for instance, where the ICC was born, the land claims agreements reached between the Inuit (and indeed all Alaska Natives) and the US government were intended to appease Alaska Native demands for a share in resource profits, while clearly subsuming their lands under the sovereign rule of the United States. The creation of Alaska Native corporations, established to manage the business activities taking place on native lands, served to steer interests primarily in the direction of development and away from political demands for alternative governance systems. In Canada, similar land claims agreements and corporations were established, yet there was a greater organization on the part of the Inuit for greater political autonomy. Probably the most far reaching success for the Canadian Inuit on this front was the creation in 1999 of the territory of Nunavut, which, like Greenland, is populated primarily by Inuit. By becoming Canada's third territory (joining Northwest Territories and Yukon), Nunavut has achieved significant self-governing powers, although still less than Canada's ten provinces. Yet, with the clear establishment of Canada's sovereignty over Nunavut in the land claims agreement, the government of Nunavut, like those established through other land claim settlements, is inherently limited to domestic economic and social issues, giving it little reason to stray from the ICC's political stand on sovereignty and statehood more broadly. In fact, ICC Canada is led by a board of directors made up of the elected leaders of the four Canadian land claims settlement regions: Inuvialuit, Nunatsiavut, Nunavik and Nunavut.

Turning back to Greenland, there are a number of reasons why this particular Inuit territory is seeking outright sovereignty. The first underlying factor for Greenland's embrace of a vision for full independence and sovereign statehood is undoubtedly geographical. Greenland, unlike the Inuit regions of Canada, the United States and Russia, is, in fact, a completely separate land mass, removed from the decision makers in Copenhagen not only by thousands of miles but by an ocean. A Canadian Inuit activist, asked about Nunavut's ambitions in relation to Greenland, mentioned this geographical dimension. He argued that Nunavut's claim to autonomy in Canada is based on maintaining a society distinct from that of the non-Inuit Canadians farther south. Claiming autonomy, then, was a possibility within the Canadian federal system. According to the respondent, however, this distinctiveness is even greater in Greenland, as it is 'an island 2,500 miles removed from the main part of the Danish society', which makes its demand for independence 'a logical proposition'. This geographical reality has long given Greenland, a rugged mountainous land whose population has survived from hunting and fishing for thousands of years, a sense of autonomy and distinction from the small, flat, farming land of Denmark.

Furthermore, the ocean separating Greenland from Denmark marks this relationship much more clearly as classically *colonial*. By contrast, in the other four Arctic littoral states, Inuit (and other indigenous peoples) inhabit territories that are more often characterized as *frontiers*, zones into which the dominant culture is gradually expanding (see Chapter 5).[1] In these states, settler populations introduced and slowly established different values and customs in the nations' Arctic peripheries, thereby diminishing to some extent the distinctiveness of the indigenous societies residing there. In Greenland, on the other hand, Denmark's attempts at fostering immigration to the distant island were only partially successful, and a large Danish presence was never established. This demographic reality, furthermore, is particularly problematic in a European context, where the nation-state has long been linked to ethnic identity, as opposed to the more civic-oriented national identities associated with the United States and Canada, for instance. Thus, even as the draconian assimilation policies of the 1950s and 1960s began to wane across the Arctic, multicultural approaches to ethnic differences in the Danish Kingdom were still largely rejected, thereby creating a larger incentive on the part of the Greenlandic Inuit to strive for outright independence.

Nonetheless, it is important to note that the independence movement in Greenland has never been formulated in ethnic terms, which has been the norm for demands made by the ICC and other Inuit groups in the establishment of land claims agreements.[2] If Greenland's pro-independence leaders were to position their right to sovereignty as the ethnic right of an indigenous people, it would raise questions regarding the status of the prospective state's non-indigenous minority, and this could end up constituting a barrier in the push for an outright break with the existing state. Instead, in Greenland the Inuit elites speak of *Greenlandic* nationalism, which immediately positions the Inuit not as an ethnic group that has special rights within the existing nation-state, but as part of a people that is distinct from the existing state that currently maintains sovereignty over them. In such a formulation, all people born in Greenland are included in the conception of a Greenlandic nation.

Lastly, and related to the above, the difference between the political leadership in Greenland and the ICC, as well as between Greenland's pro-independence activists and the Inuit leadership in Alaska, Canada and Russia more generally, points to a simple divergence in opportunity. A significant reason for the ICC's open skepticism towards the state system is that this system potentially limits the power of the Inuit if they are regarded as ordinary citizens without special claims. Hence the ICC has sought to downplay the power of the nation-state in relation to its indigenous inhabitants. Again, due to the geography and the nature of national identity in the United States and Canada, the aim of full independence has been a non-starter, and consequently Inuit leaders in these countries have sought to make compromises in order not to be cut out of decision making all together. This is even more the case in Russia, where the state has been loath to grant autonomous political power to indigenous peoples. As a part of this compromise, indigenous peoples in the United States, Canada and Russia have stopped short of calling for an independent, sovereign homeland (either within or across state borders) and instead have generally embraced American, Canadian or Russian citizenship, even while attempting to carve out a level of ethnic and territorial autonomy. In Greenland, on the other hand, due to its unique geographical situation and historical trajectory, Denmark has been exceedingly cooperative, with public statements made by government officials that Denmark will not stand in the way of the Greenlanders' will. The opportunity in Greenland for a full break with the colonial past, and the achievement of gaining complete control over both domestic and foreign affairs, appears not only desirable but tangible.

Greenland's emerging embrace, not only of independence but also of a state-centered political organization of the Arctic, is reflected in its position with regard to the Ilulissat deliberations and the continued discussions being maintained by the five Arctic Ocean littoral states. It was, in fact, the Danes who were the architects behind the Ilulissat Declaration, organizing the meeting in Greenland. For Denmark, the move to establish the declaration was, officially, to put to rest any uncertainty about the Arctic. The massive race for the Arctic as it was being presented in the press was to be averted. In the process, of course, Denmark managed to establish itself as a diplomatic power broker in the Arctic, and create a place for itself in the elite club of the Arctic Five, even though, given the drive toward Greenlandic independence, its claim to being an Arctic Ocean coastal state would be endangered if Greenland were to break away.

As was mentioned in Chapter 1, the Ilulissat Declaration was frowned on by the various players who were not invited to the table, including the ICC and the five other Arctic Council permanent participants representing indigenous peoples. The ICC in particular was irked by the declaration as it has historically attempted to downplay the role of the 'sovereign' nation-state, emphasizing instead the state-transcending identity of the Inuit people. In the Greenlandic government, however, there was a fundamental shift away from the ICC's position. As one official interviewed stated:

> Seen from up here, we believe that the decisions made about our region should be made by the players within this region . . . when it comes to things that strictly concern the neighbors to this area, or us living here, well then . . . . It's for the countries in the area to decide that.

The focus placed on 'countries' is telling here, as it neatly circumvents the role of non-state actors that generally is emphasized by Inuit organizations.

For these reasons, Greenland did not oppose the Danish-sponsored deliberations to convene the Arctic Five meeting. As one Government of Greenland official stated: 'The Arctic Five is fine with us, because we were part of it. And the invitation to the meeting (Illulisat) was a joint invitation . . . to the Danish Minister of Foreign Affairs and our Premier'.

It is interesting to note that after the Ilulissat Declaration Copenhagen began to distance itself from the Arctic Five, apparently in response to the barrage of criticism that it was receiving from those who had been left out. Thus, when Canada invited Denmark to the second Arctic Five meeting in Chelsea, Québec, the Danes gave a tepid agreement to attend, but then the new Danish Foreign Minister, Lene Espersen, ultimately cancelled her participation in the ministerial meeting as it conflicted with her family vacation plans in Spain. Denmark was the only country not to send its foreign minister, for which Espersen drew considerable domestic flack. Kuppik Kleist, the Greenlandic premier at the time, did attend (although apparently he was not explicitly invited). Among Greenlandic officials the absence of

4.2   Representatives to the Five Ministers Meeting, Chelsea, Québec.

Espersen was seen as a lack of commitment to the Arctic, thus positioning the Greenlandic government as more open to the Arctic Five than the Danish government. Whether Kleist's sole presence boosted Greenland's legitimacy as a soon-to-be-independent state is difficult to say. For Greenlanders, however, the meeting was seen as a success, as Kleist's presence marked the premier as more than the head of a self-governance arrangement; in their eyes it put him at the table as a world leader. Yet the confidence in such a feeling could not be fully embraced due to the way the meeting was adjourned – with a photo op that included Denmark's Minister of Justice Lars Barfoed, who attended in place of Espersen, but 'somehow' excluded Kleist. Greenland's ouster from the staged media moment proved to be a reality check: as things stood, the Greenlandic premier could rub elbows with the world's leaders but he certainly was not yet seen as the leader of a sovereign state.

The Greenlandic government's attitude toward the Ilulissat and Chelsea meetings suggests that it has embraced the Arctic Five as a legitimate forum, in which the nation-state stands as the ultimate template to organize political space. This is not a surprise if we consider that the three leading political parties in Greenland all have as their ultimate objective independence from Denmark; that is, the creation of a new state. Yet, such an approach has not gone without criticism. For instance, Canadian scholars Menno Boldt and Anthony Long have argued that the adoption of a state system solution to indigenous interests only serves to further cement Western power structures and dominance. In a similar vein, the ICC holds that a more effective strategy for promoting Inuit empowerment has at its core the advocacy of a non-state-oriented circumpolarity. This strategy, according to ICC leadership, has helped the Inuit in advancing their agendas, which means not only gaining a greater say at the table of power but also becoming accepted and empowered as a people that exists outside the dominant Western political, economic and even spiritual worldview.

It is in this argument that the distinction between the cultural politics of the Greenlandic government and that of non-state-centered Inuit organizations comes to the fore. The Greenlandic government is pursuing a nationalism, or Greenlandization, that is distinct from, although not necessarily exclusionary of, a broader Inuit identity politics. Greenlandization is here more a political project to gain access to the realm of high-level decision making, rather than a project to preserve or empower a particular culture. In this way, self-rule and future independence is based on a discourse of public government, not on ethnic identity. As a Greenlandic government official explained,

> When Greenland is sitting at the table ... it represents the public government of Greenland, and it's not just an indigenous government, it's a public government. De facto it is very much an indigenous government, but formally it is not, it's a public government.

In this way the Inuit of Greenland have managed a different political trajectory than the more ethnic and culturally focused struggles for self-determination that have been pursued in, for instance, Alaska and Canada. The Greenlandic political efforts are squarely focused on achieving independence on the basis of anti-colonial, separation nationalism.

Yet the 'independence at any price' focus in Greenland, which, although unique, still could serve to inspire future indigenous movements, needs to be put into perspective. What price is being considered here? And what is the timeframe before full independence can be achieved? To those most familiar with Greenland's current economic situation, there is much cause for caution. Greenlandic government officials all point to the long road ahead. And in Denmark, government officials do not see the numbers adding up – at least not yet. As one Danish government official responsible for Greenland relations stated:

> It's as if everybody sees self-government as, you know, the first foot down, and then the next foot inevitably must be independence. It's a country in its own right. It has its own language, its own people, its

own borders. Of course it's an underlying discussion: 'What would it take for us to become independent?' But that doesn't mean that independence is imminent, because it simply would not be good for the people of Greenland.

Oil: Boon or Cultural Suicide?

The above quote begs the question: When would independence be good for Greenland? Answering this question can also help us gauge the viability of other Arctic independence movements, even if they are not yet fully on the horizon. Currently, the main issue regarding an independent state for Greenland is the question of its feasibility, that is whether it would actually be able to command sovereign rule over its vast expanse. To throw light on this question we can consider a number of criteria for the successful execution of sovereign rule as initially laid out by international relations theorist Stephen Krasner. Krasner identifies four necessary components of sovereignty: (1) the ability to 'regulate the movement of goods, capital, people and ideas' across national borders; (2) the capacity to exert effective domestic control over sovereign territory; (3) gaining legal recognition by other states; and (4) the capability to function autonomously, without 'authoritative external influences'.

At first glance, Greenland appears relatively well positioned to claim sovereign control via independence. The outside world, for instance, has no reason to refuse Greenland its independence, so long as the Danish parliament approves it. In terms of the movement of goods, capital, people and ideas Greenland already exerts significant power, as evidenced by its decision not to be a party to EU (formerly ECC) rules and regulations despite Denmark's membership. Similarly, Greenland has been working to impose its own distinctive immigration rules, intended to meet the specific needs of Greenland's labor market. Clearly, with independence, the Greenlandic government's ability to execute such decisions would only be enhanced. Finally, domestic control over Greenland is primarily challenged by the costs of distance between its remote inhabited enclaves, yet it also has the distinct advantage of maintaining a largely homogenous population that offers very little resistance to the legitimacy of the Greenlandic government.

Nevertheless, it is Krasner's fourth point that is most problematic – namely, the ability of the state to function autonomously. Like any state, Greenland's ability to rule over its territory is hampered by the inevitable compromises made in the process of plugging the national economy into the global flow of goods and capital. Yet in Greenland this compromise is exacerbated by its very small population and the underlying weakness of its economy. It is explicitly clear to the Greenland self-rule government, as it is to the outside world, that any independence from Denmark is predicated on the discovery and subsequent extraction of oil and gas off the Greenlandic coast.

Currently Greenland is dependent on Denmark to bankroll the running of its government. Greenland receives a block grant of 3.2 billion Danish Kroner (US$550m), which makes up about one-fourth of its entire gross domestic product. As one member of the current Greenlandic government put it,

> In order to make that kind of money ourselves we need to have more activity in this country ... we need to work more, we need to build more, we need to get more industrialized .... The opportunities we have are mineral resources. And possibly, oil resources and gas resources.

Put more bluntly, a Danish government official said,

> I mean the only way they can survive is to find some oil. It's not nice to put it like that. But that's the hard realities .... Develop the hell out of it, see what you can get.

Yet, becoming an oil and gas exporting nation would mean a massive restructuring of Greenland's economy, which in turn would have to be driven by external competencies. Foreign oil companies and the states in which they are headquartered would likely gain an increasing stake, and subsequent influence, in Greenlandic affairs. Comparisons, here, with oil-rich countries in the Middle East abound, with references made to a Kuwaiti or Qatari model in the Arctic. Among many of those interviewed, such a possibility was spoken about cynically, lamenting the prospective loss of what makes Greenland unique.

Certainly, traditional Inuit culture, based on remote living and subsistence hunting, would become significantly more difficult to sustain if Greenlanders were to 'develop the hell' out of their territory in search of an oil-based independence, and the same challenges would be faced by any other Inuit region seeking independence. In fact, many Inuit territories already have embraced resource extraction and transnational capital penetration. As one high-ranking Greenlandic ICC official noted,

> One opportunity also creates several problems, like the social consequences of what we're doing ... like we have seen and we are told by all the others from Alaska, from Canada, from Chukotka. Look and see what happened with industrialization.

In Greenland the move to industrialization, driven by a booming oil export industry, would inevitably mean a greater need for Inuit populations to concentrate within towns close to the oil and gas infrastructure, while still requiring substantial immigration to provide the labor and competence to run a sophisticated resource-extraction economy. The Inuit would be significantly diluted as a percentage of the Greenlandic population while at the same time being absorbed into a completely different economic structure. In other words, an 'Inuit' petro-state would in all likelihood lead very far away from the cultural integrity and self-determination that is largely at the root of the ICC's political framework, as well as wiping away the traditional Greenland that lies close to the hearts of the Danes.

The standard response from the Greenlandic government to such concerns is that the cultural identity on which the ICC hangs its political hat is in many ways a myth that conceals a rapidly changing Inuit reality. In Greenland, for example, about 75 percent of the population already live in urban settlements. Hunting and fishing is, for the most part, a pastime, although a culturally significant one. One Greenlandic official estimated that fewer than 5 percent of Greenlanders actually live a subsistence hunting/fishing lifestyle.

The prospect of a large aluminum smelter in Maniitsoq, Greenland, run by Alcoa is a good example of the contested nature of Greenland's development future. As a representative of the Greenlandic government explained in an interview:

> Now, of course, Alcoa is coming in and probably buying half of the country, or the whole thing .... The old mayor in that town where the aluminum smelter is going to be, he was confronted with the question whether there should be people coming from the outside to work there. Yes, but they would have to speak Greenlandic he says. And that is one of the things that remains to be seen.

Another official was more optimistic: 'If Alcoa comes of course this will have an impact, but I think that our resilience is so powerful that we can still maintain a great deal of our culture.' When pressed to elaborate, the official explained his idea of promoting seasonal work, with laborers working in factories and oil platforms and then returning to the villages to live a more traditional life. In other words, full urbanization would not be encouraged:

> We can maintain a number of villagers working in mines, in cities, further south in oil exploration, and then they can go back to live their life, to fishing and hunting, join the family, why not? Culture is not necessarily what you do to maintain a living; it is also what you do after

four o' clock. We want to maintain the basics of our way of doing things. This is our idea of how to do it.

Thus, Greenlandic officials speak of the importance of cherishing and maintaining their culture, yet they stop well short of claiming that their culture is in any way incompatible with Western norms, a sentiment that can indeed be heard within the ICC and other indigenous organizations as well.

With regard to the environment, the Greenlandic government has also taken a very different approach to that usually espoused by Inuit non-governmental organizations. While Inuit leaders in Alaska and Canada tend to express a deep concern over the effects of transnational corporate involvement in the Arctic, especially as it pertains to the maintenance of traditional lifestyles, the situation is different in Greenland, where the Inuit-dominated government has achieved practically full control over resource development and its profits. Yet, as Greenland continues to play out its ambitions as an aspiring sovereign state, we end up with the awkward event in which Greenland's prime minister is seen publicly scolding Greenpeace for interfering with oil exploration efforts off the Greenland coast. Indeed, Greenland's government now views global warming as a potential boon to its development chances. As one key Greenlandic official noted, some economic endeavors are 'getting more and more interesting because of retreating ice. So in that sense you can say [climate change] is positive if you get activities that can give this country more revenues and jobs'. From this perspective, as the ice melts, resources are exposed, shipping lanes open up and agriculture becomes a possibility in the south.

The idea that Greenland should be denied its right to exploit its own resources, furthermore, is met with great suspicion on the part of the Greenlandic government. As one official succinctly stated, 'Other countries that have those resources are using these resources. Why shouldn't we?' And indeed Greenlanders are, or at least they are trying to. While in Nuuk conducting interviews, there was a veritable storm of activity as licenses were being awarded to new companies and companies with existing licenses were moving their drilling rigs into position, reporting back on a regular basis on their progress. In the midst of this activity one government official, who was constantly being interrupted by phone calls from the Greenland Bureau of Minerals and Petroleum and the prime minister, said:

I think Greenland will not look the same in ten years time. So it's up to us to handle anything in the best way we ever can, without – I would say – destroying the culture and the environment and the people here.

This is no doubt a tall task, and it seems there is currently no consensus as to how best to achieve it. Perhaps even more problematic, however, is that there does not appear to be a real public discussion going on regarding the challenges.

Part of the problem is the weakness of organized political pressure groups in Greenland, particularly from the environmental side. Although alliances between indigenous and environmental groups have achieved some tangible successes, most notably surrounding the Arctic Council's work on persistent organic pollutants in the Arctic atmosphere and the health risks that these pose to the region's residents, in many other instances relations between indigenous and environmental groups have been strained. Addressing the situation in Greenland specifically, one Greenpeace official explained,

> In order to really influence politicians you need to have a strong constituency in Greenland. It doesn't help that we have a strong foothold in Denmark as long as we don't have it in Greenland – and there are many reasons why we don't have it.

In fact the environmental movement is particularly weak in Greenland. The main reason for this is that Greenlanders associate environmental activism with struggles over what they see as the living resources of the sea – whales and seals. These are the issues most dear to them and they have all too often seen themselves on the defensive end of these environmental campaigns. A problem for US and European environmental movements that are seeking to have an impact on the Arctic is that Western environmental non-governmental organizations like Greenpeace or the WWF tend to work on a broad range of issues. Thus, as potentially popular campaigns like those against global warming and toxic waste disposal are bundled with issues like seal hunting, coalition building becomes increasingly difficult (see Chapter 7).

Yet it was not only NGOs that lamented the lack of a critical voice in Greenland: government officials did as well. There was a concern that the people themselves were walking into the future with their eyes not open wide enough. It was as if the government did not want to push Greenland into a drastically new future without at least some internal discussion. One government official interviewed explained, almost with a hint of embarrassment, how Alcoa arrived in Greenland to present its aluminum smelter project. The Alcoa representatives had expected to be greeted by a host of critical NGOs, posing difficult questions. Instead they were greeted by friendly citizens waving Greenlandic flags; there was no deeper questioning of what a massive aluminum smelter with a largely foreign labor force would mean for the region. As another official expounded,

On these major industrial projects ... we still need to develop our democratic hearing processes. A lot. We have meetings, public meetings, and we have processes where we hear different parties and groups, but we haven't properly made a process that satisfies all, and where you make sure you don't just hear, but perhaps also take consequence from that which is actually being said. So that is a major challenge.

Indeed, despite being at the forefront of indigenous autonomy and perhaps independence in the Arctic, it is a steep learning curve for the Greenlanders, not only to become a self-sustaining society but also one that can govern itself democratically. The challenge in Greenland is especially high due to the seductive lure of independence. As opposed to Inuit areas in Alaska, Canada and Russia, where development has often been tied to external interests and hence is often viewed with suspicion by local residents, in Greenland, where domestic authorities have more say and more to gain, the impetus to think critically is inevitably blunted.

Allowing the Break-up

As Greenlanders look towards a future of greater development and hence greater economic autonomy from Denmark, a lingering question remains regarding how Denmark views its changing relation to Greenland. In the self-government agreement, Greenland remains a part of the Danish Kingdom, subject to the Danish crown and Danish foreign policy. For the time being an agreement has been reached regarding potential oil revenues – 50 percent of any revenues will be used to pay off the block grant. However, it was also decided that the agreement over oil income would have to be renegotiated if the amount ends up being so large that the entire block grant becomes unnecessary. Would Denmark, being so close to becoming a wealthy oil nation, allow Greenland to declare independence? Given the media's continuous depiction of a state-based free-for-all over the Arctic's assumed wealth, most outsiders would find it hard to believe that Denmark would willingly give up its stake in the High North. However, as we have suggested throughout this book, cooperation and diplomacy are the norm in the Arctic, and this tendency applies here as well. In fact, all indications are that Denmark would indeed grant Greenland its independence if it were requested.

Again there is a clear difference here between Greenland's situation and that of other Inuit territories. Due to its particular self-presentation as a distant colony, Greenland has long positioned itself as distinct from Denmark, and hence as having the right to pursue national self-determination. In the United States, Canada and Russia, in contrast, the indigenous populations of the North are viewed as citizens with special claims, but any talk of independence is quickly dismissed as unrealistic, radical and off the table. Yet this is not to say that the Danes, who are most directly confronted with the prospect of a breakaway Inuit state, have fully come to terms with the idea. Even here, where the legitimacy of Greenland's self-determination ambitions are recognized, many Danes still find it difficult to accept a Denmark without Greenland.

Speaking to a Danish parliamentarian representing Greenland (Greenland has two representatives in Denmark's parliament) the facts were clearly spelled out: 'Denmark is not part of the Arctic, it's only a part of the Arctic because we are; Greenland is in the Arctic.' This statement, however, can have multiple insinuations. On the one hand, it could mean that Denmark is an odd player in the Arctic and that its role there should defer to Greenland, including giving Greenland independence if it is requested (this is undoubtedly the meaning intended by the speaker). On the other hand, it could mean that Greenland is Denmark's only 'ticket' to a region that is becoming increasingly important in terms of resources, shipping routes and geopolitics; in other words, Denmark has a strong self-interest in maintaining formal ties to Greenland. As it turns out, there is an element of truth in both these readings.

In many respects, the Danes are still coming to terms with the self-rule arrangement, agreed to by the Danish parliament and implemented in 2009, which has given Greenland extensive rights. The increasing powerlessness felt on the part of the Danes is especially noticeable with regard to Greenland's newfound right to subsoil resources. This issue has come very much to the fore since the Gulf of Mexico Deepwater Horizon incident, when there were calls across the world to put a moratorium on offshore oil drilling. One Danish government official explained how these concerns over oil drilling were playing out with regard to Greenland:

Of course a lot of people, even politicians, say, 'Oh how is that? I think it's dangerous, you should forbid it, I mean why don't you stop it.' And we could say, 'Why don't I stop things in Afghanistan, and Russia, and Malaysia, and things like that? It's not my business.' 'Yes you should ban this, and how can you as minister responsible be in favor of that?' And she has to say, 'Sorry but this is something which is decided by the Greenland Self-Government. And I have nothing to add.' ... You can read between the lines in the newspapers, that it was, well, it was also the first time that she really had to say, 'Well I've given it away. It's not my business anymore.'

This self-reflection and pondering over lost power was not limited to environmental concerns. Signing over all the rights to Greenland's mineral and hydrocarbon wealth was not viewed as a fair and responsible thing to do by all Danes. To some, particularly supporters of the right-leaning Danish People's Party, the Danes had negotiated a bad deal. As one party leader explained,

> We wanted a fair split. Or maybe, we suggested, also at some point . . . that we would create like the Norwegians, an oil fund that would be just like a bankbook collecting wealth that could then be used mainly to the benefit of the Greenlanders, but also being used by Denmark. We compared it to this, where the Canadians gave the Inuit 5 percent. We suggested 50 and then end up giving them it all. I find the whole [deal] unfair. And so did the rest of the population here. I can assure you they were on my side.

On his latter point there is disagreement. One government official, not from the Danish People's Party, responded directly to this claim, 'I think most people feel that it would be thievery to insist on hanging on to the rights of the underground'.

Either way, it would nonetheless be wrong to depict Denmark's opinions about Greenland as being based entirely on a desire for power in the Arctic or a greed for the oil wealth that Greenland may be sitting on. In responding to the question of Denmark's relation to Greenland in the various interviews with Danish government officials and politicians, a number of themes emerged. To begin, it was repeatedly emphasized that Greenland is, in fact, by law, still a part of the Kingdom of Denmark. This is important to Danes, as their realm, their sovereign space, has for generations included Greenland as an integral part. It is taught in the schools and it is celebrated at Christmas time. The royal family in Denmark has close ties to Greenland, with Crown Prince Frederik having participated in the elite Navy dog sledge patrol of northern Greenland – the Syrius patrol – and now recently having christened his twins with traditional Greenlandic names. This love affair with Greenland, furthermore, is said to go both ways. The Greenlanders are avid royalists who follow the doings of Danish royalty closely.

Denmark's sovereign claim, its place on the map in the Arctic, is thus deeply embedded in the Danish consciousness: it is part of the national identity. Much of this is historical, a point frequently made in the interviews: the various Danish Arctic explorers, the glory of Knud Rasmussen, the prolonged presence of Danish settlers in Greenland. All of this was mentioned as a way to downplay the idea that Denmark, a flat, temperate land of farmers, is an awkward Arctic player. Part of this history, furthermore, has led to family bonds that now, in the view of many Danes, inextricably tie the Danes to the Greenlanders. One Danish politician went even further in linking the two peoples, stating, 'I don't think there is any family in Greenland where they haven't got Danish blood; it's so intermingled after all these years we have been together.' This togetherness, furthermore, has been viewed by most Danes as a benign paternalism on the part of the Danish Kingdom. Denmark's Arctic policy has always squarely focused on the human dimension, pursuing a welfare policy in Greenland that was intended to emulate the well-being found in Denmark.

Yet in speaking to Inuit Greenlanders, especially in the government, their feelings appeared more reserved and lacked the connection to the mother nation found among the Inuit of Canada and the United States. As one Greenlandic official explained,

> We are Danish citizens, but ... well, there is something not completely right about that. Because you will find Inuit, you will find Greenlanders that don't speak Danish, that have never been in Denmark but are Danish citizens, who don't identify with Denmark, who don't identify with the Danish flag, and the language and everything. And there is a problem in that.

Or put even more succinctly, one Greenlandic representative to the Danish parliament declared, 'I would not be a Dane, I'm Greenlandic first.'

This discrepancy in attitude does not bode well for the Danes' hope of hanging on to their beloved Greenland, especially if oil were to be discovered there. But this is not to say that the Danes would not let Greenland go: indeed it is accepted across the political spectrum that if so desired the Greenlanders could have their independence, if they truly desired it. As one of the Danish government's top officials in Greenland stated matter of factly, 'If the people of Greenland were to decide in a referendum that they wanted independence, they have it. I mean it's not for Denmark to decide'. Or put differently by a parliamentarian from the Danish People's Party, 'If the Greenlanders really, really want their independence, then of course we should not deny them that right. That's the right of every nation to choose'. Nonetheless, there is a disbelief in Denmark that Greenlanders would want independence. In this sense, a Danish granting of independence would first and foremost be on the basis of accepted geopolitical standards of the day; in other words, it is not in the Danish self-image to resist the national self-determination of a people living thousands of miles away. Thus, if the demand were unanimous, independence would be granted but it would not be done easily. As one Danish official confided, 'If ever Greenland were to declare its independence, it would make an enormous big wound in our hearts ... I couldn't imagine a Denmark without Greenland'.

Geopolitics

Greenlandic independence would also have significant geopolitical implications, for the Arctic region and beyond. Clearly, independence hinges on massive development in the region, primarily around the oil and gas sector. As a result, the region would undoubtedly gain significantly in geopolitical importance, and Greenland, with its small population, would have to rely almost completely on foreign states for security and search and rescue operations. Practically all Greenlandic officials agree that Greenland would remain firmly in NATO. It is possible, furthermore, that the US Air Force base in Thule would gain renewed strategic importance and may be accompanied by further US military installations. In addition, the influx of petro-dollars would plug Greenland into the financial circuit of capital centered on the United States, and may even bring pressure on Greenland to adopt the dollar as the country's official currency. With increasing American contacts it is quite possible that Greenlandic children would study English rather than Danish.

With such thoughts in mind, there is much speculation that Greenland, which has generally been viewed as Europe's westernmost edge, could shift to the US sphere of influence and become rather the easternmost edge of North America. As one key Danish official explained,

> I do not believe that either Russia or the United States would allow for a sort of geostrategic vacuum in the Arctic, and I am convinced that perhaps the US would be the most probable candidate. It would be very, very fast to do a Puerto Rico thing on Greenland if Greenland were to become independent.

The fact that the United States at one point had offered to purchase Greenland from the Danes should not be forgotten.

Greenlandic officials, in turn, have openly expressed their interest in US partnerships, much to the chagrin of some Danes. The United States has reciprocated this flirtation, which has led to some frictions with Denmark. Referring to the invitation of a Greenlandic delegation to Washington, DC, one Danish politician declared,

It's normal practice if you want to send an invitation to the Greenland authority, it goes through the foreign office here [in Copenhagen]. And I don't think that the Americans have followed that practice .... This inviting the Greenland leaders to Washington and receiving them ... almost like an independent nation, I don't think that's very fair.

Clearly there is a sense of competition regarding Greenland. Whereas the United States seeks to get into the good graces of the Greenlandic elite, the Danes still demonstratively declare their sovereignty over the island.

The Danes have also increased their military presence in the Arctic, as have other nations in the region. They have recently established an Arctic command in Nuuk, from which they now patrol Greenland's coastal waters. Officials at Denmark's Ministry of Defence explained that their activities in the Arctic were aimed at securing Denmark's' sovereign claim in the region. Yet, as was frequently pointed out, the activities were in no way to be viewed as aggressive grandstanding: rather, they were presented as a service to Greenland that is understood to fall under the obligations that Denmark has to its Greenlandic citizens. In concrete terms this means primarily aiding in search and rescue. Hence, in the eyes of the Danish military it is simply doing its constabulary duty in keeping its sovereign realm safe, and in the process this sovereign realm is reaffirmed as being Danish. As one politician firmly stated, 'For the time being, there's no doubt about who's in control. We are.'

Yet again, the question is: For how much longer? The military presence around Greenland is certainly not an attempt to claim Greenland unilaterally. When the same official from the Ministry of Defence was asked about the possibility of Greenlandic independence he stated, 'That's for them to choose. Until that, we are just trying to provide them with the best conditions for making that choice.' There was not the slightest hint of resentment in the official's voice when saying these words. For him, the Danish military was simply doing its duty; politics was for someone else to decide.

Turning back to Greenland, it is interesting to note that the current government refuses to feel any obligation in the face of this service that Denmark provides. Rather, the Greenlandic government presents itself as an autonomous actor that will work with whomever it sees fit to pursue its interests. As one official stated,

When Greenland becomes independent, it is not a question of us saying we will not cooperate with Denmark, we will not cooperate with the US, we will not cooperate with Canada. We will cooperate on many areas .... Some issues we take bilaterally, and other things need to be multilateral.

In this sense, the vision is one of true freedom. Yet, and this is an important point to note for any indigenous movement seeking to emulate Greenland, political geographers have long pointed to the diminishing autonomy that small, economically weak states around the world experience, and there is no reason to believe that Greenland would be an exception. In a globalized world market, within which Greenland wants to firmly plant itself, there will be numerous compromises to make.

These compromises will center on how Greenland answers a key question: To which side of the Atlantic will it tilt? Ultimately this is a choice between two major projects: the supranational project of the European Union or the continued project of the United States as a global, unipolar leader. Sacrifices to traditional culture and traditions are already being made in the push for development. As this development continues, Greenland will have to decide with whom to work most closely. A cherry-picking approach is unlikely to work as economic, political and cultural realms of cooperation will overlap and reinforce each other. In this sense, soon after independence, Greenlanders will undoubtedly find themselves in a similar position to many newly independent colonies throughout history: they will have a seat at the table but their trading conditions, their currency, their finances, their security and their language will be tied up in the larger global power structures that currently are in place. In the meantime, nothing of significance in terms of Arctic governance will have changed; there will have been a redrawing of borders and a switch of actors and flags that are now present at the negotiating table but the Arctic will still be a space dominated by the modern imaginary of state jurisdictions, and these states will be intertwined with big capital interests maneuvering to access the unexploited wealth of the North. Greenland will have joined the status quo, with increased international recognition but at a domestic price that is still difficult to foresee.

To conclude, then, it can be said that the Greenlandic leadership is taking a gamble, seeking to maximize empowerment by playing the dominant political game in search of statehood. The level of sovereignty achieved, and the toll of the compromises demanded to achieve it, remain to be seen. In the meantime the Greenlanders are actually reinforcing a modern political organization of space in the Arctic, turning their back on the more questioning stance of the ICC and other indigenous organizations in the North. In their tacit endorsement of the modern imaginary, however, Greenlanders are proposing a future that would reproduce not just the principle of formally equivalent sovereign states but also the reality of a world in which there are significant differences in power between these states, differences that are reinforced by the seemingly hard shells of territorial borders. Perhaps, with time, such realizations could lead to the creation of a state that actually becomes a dominant voice in the push towards a more empowered and diverse multilateralism in the Arctic, and around the world – a state that actively seeks to rethink the status of the state in a world in which people, ideas, goods and capital cross borders as well as being contained by them. For now, however, independence and statehood at any price is the mantra in Greenland, as other Arctic indigenous groups in the wings watch in anticipation and take notes.

5.1   China's Yellow River Research Station, Svalbard, Norway.

CHAPTER 5

RESOURCE FRONTIER

Much has been made of the Russian flag that was planted in 2007 on the seabed at the North Pole (see Chapter 2). However, another flag was placed in the Arctic six years earlier that drew much less media attention. On 31 October 2001, a government-sponsored team of scientists led by Gao Dengyi raised the Chinese flag to mark the opening of the China Yilite-Mornring Arctic Scientific Expedition and Research Station in Longyearbyen, Svalbard, Norway – this only two years after the first Chinese state-led trek to the North Pole. Although the station was designed to support research on 'climate, environment and life-forms', state media also blithely mentioned 'resources' among the mission's scientific pursuits. Commenting on the Arctic states' responses to the growing Chinese interest in the North, one Canadian official noted, 'Arctic states have said, "Hold it, that's our backyard and we're not necessarily sure if you'll get to be at the table when we divvy up the resources and you don't have any prior claims."' The 2001 Chinese flag hoisting may have been in the service of Arctic science, but a prominent individual in the US Arctic research establishment suggested in an interview that the entry of Chinese (as well as Japanese and South Korean) scientists into the region was not without motive:

I think what you're seeing here is establishing their presence up there. So when there's any kind of future decisions made on who's got jurisdiction, who has access to the resources, they can say they have been up there.

Thus, claims can be made in the Arctic and eventually to its oil without 'owning' Arctic lands and waters or having UNCLOS-sanctioned exclusive rights to its resources. If there is a race for Arctic natural resource riches, there are many more than the five littoral Arctic Ocean states in the race.

Throughout discussions with Arctic government officials, NGO leaders and industry representatives, some believed that the Arctic's resources were up for grabs by states and corporations, and even, in a manner of speaking, by environmental organizations (see Chapter 7). In this imaginary, the Arctic is seen as a global commons that holds valuable minerals, marine and terrestrial wildlife and, in particular, oil and gas. Central to this vision is an explicit or implied notion of the Arctic as a frontier where ownership of the resources is somehow undetermined, and that these resources can be had by any entity that is quick enough to obtain them. Over the course of our interviews, the Arctic was referred to as 'a storehouse of natural resources', 'a big bazaar', 'America's resource province' and 'a frontier for our resource sector'. Yet beyond the basic presumption that the Arctic contains resources that can be exploited, the details of the Arctic as a frontier commons and the opportunities (and threats) that this vision implies are much more differentiated.

If the commons includes lands or waters that are 'common' property, this means they are neither privately nor publicly owned. Often, a commons is thought to be immanently threatened by what ecologist Garret Hardin described in 1968 as the 'tragedy of the commons'. The pitfall that leads to 'tragedy' is that without public or private ownership over-exploitation is inevitable. During our interviews, imaginations of the Arctic as a commons were frequently coupled with depictions of the Arctic as a frontier, a newly accessible region beyond established settlements where pioneers are free to exploit natural riches.

Although the 'pioneers' exploiting fisheries, minerals, gas and oil often hail from Arctic states, other non-Arctic stakeholders are also represented as poised to join the 'gold rush'. Extra-Arctic states such as China, Japan and South Korea as well as the European Union have made moves to establish an Arctic presence, despite the fact that the vast majority of the Arctic's commercially exploitable resources are under the jurisdiction of one or another of the Arctic Five. Furthermore, through bids to drill offshore in the Arctic states, global oil corporations – for instance Shell, ExxonMobil and ConocoPhillips – are also implicated as eager to enter more deeply into the region. The prospect of 'newcomers' gaining access to the Arctic is made all the more pressing by the ambiguous geography referred to by the word 'Arctic', which includes an ocean but also a seasonally frozen terrestrial region. This, in turn, engenders misunderstandings regarding sovereignty and rights of exploitation. In terms of the totality of Arctic land and water 'acreage', most of the Arctic – or at least its resources – are under the sovereign jurisdiction of Arctic rim states. However, as many officials, researchers, consultants and corporate entities interviewed noted, this regime governing the Arctic is often overridden by a media driven vision of the North as a resource prize that will be divvied up by whoever reaches it first.

Understandably, this Arctic imaginary as a resource-rich frontier commons faces much resistance. In particular, government officials from Arctic states are quick to correct any lack of clarity with regard to who owns the Arctic. In a basic sense of state sovereignty (in line with the Ilulissat vision presented in Chapter 1), Arctic land-based natural resources are, with the single exception of Hans Island, which is disputed between Canada and Denmark, already unproblematically claimed. Furthermore, since four of the five Arctic Ocean coastal states have ratified UNCLOS, and the United States has committed to abide without signing, the extraction of resources from Arctic states' territorial waters, 200 nautical mile EEZs and outer continental shelves out to 350 nautical miles or more is under the jurisdiction of respective coastal states. Thus, the idea of the Arctic as a global commons, frontier or not, is rejected by the Arctic Five. Nonetheless, despite its near universal dismissal by parties actively involved in, or contemplating activity in, the region, the 'resource frontier' imaginary continues to influence the debate surrounding the future of the Arctic, much like the similarly discredited *terra nullius* imaginary discussed in Chapter 2.

While the Arctic states share a political resistance to the power of this imaginary, they are not culturally or politically homogenous. From Denmark at one extreme to Russia at the other, there are substantial differences in the level of governance over resource issues granted to regional authorities. Additionally, Arctic indigenous peoples as well as non-indigenous residents have political and historic rights to their 'backyards' and, despite naive perspectives that see indigenous residents as seeking only to maintain a subsistence lifestyle, they too have a financial interest in maintaining and in some cases obtaining control of what they see as their natural resources (addressed in depth in Chapter 4, with reference to the Greenlandic independence movement). Thus, the race to develop resources is not a 'winner-takes-all' finale but a process by which many groups hope to benefit.

Resources: Types, Extremes, Challenges

If one phrase throughout these interviews could be selected as the most common description of the Arctic, it would be 'vast natural resources'. In particular, representatives of resource-extraction interests offered detailed empirical evidence through statistics and maps that supported their claims regarding the Arctic's immense natural wealth. But those who were not directly involved with this sector also clearly recognized that the globalizing interest in the extraction of Arctic resources is impacting the extended region.

PETROLEUM POTENTIAL OF ASSESSMENT UNITS AND PROVINCES IN THE CIRCUM-ARCTIC

In the Circum-Arctic Resource Appraisal (CARA), 33 provinces were examined, of which 25 were judged to have a 10-percent or greater probability of at least one significant undiscovered petroleum accumulation in any constituent assessment unit (AU) and were therefore quantitatively assessed. Shown in these three maps are the relative probabilities for all assessment units assessed and the estimated relative potentials for undiscovered oil and gas in the assessed provinces.

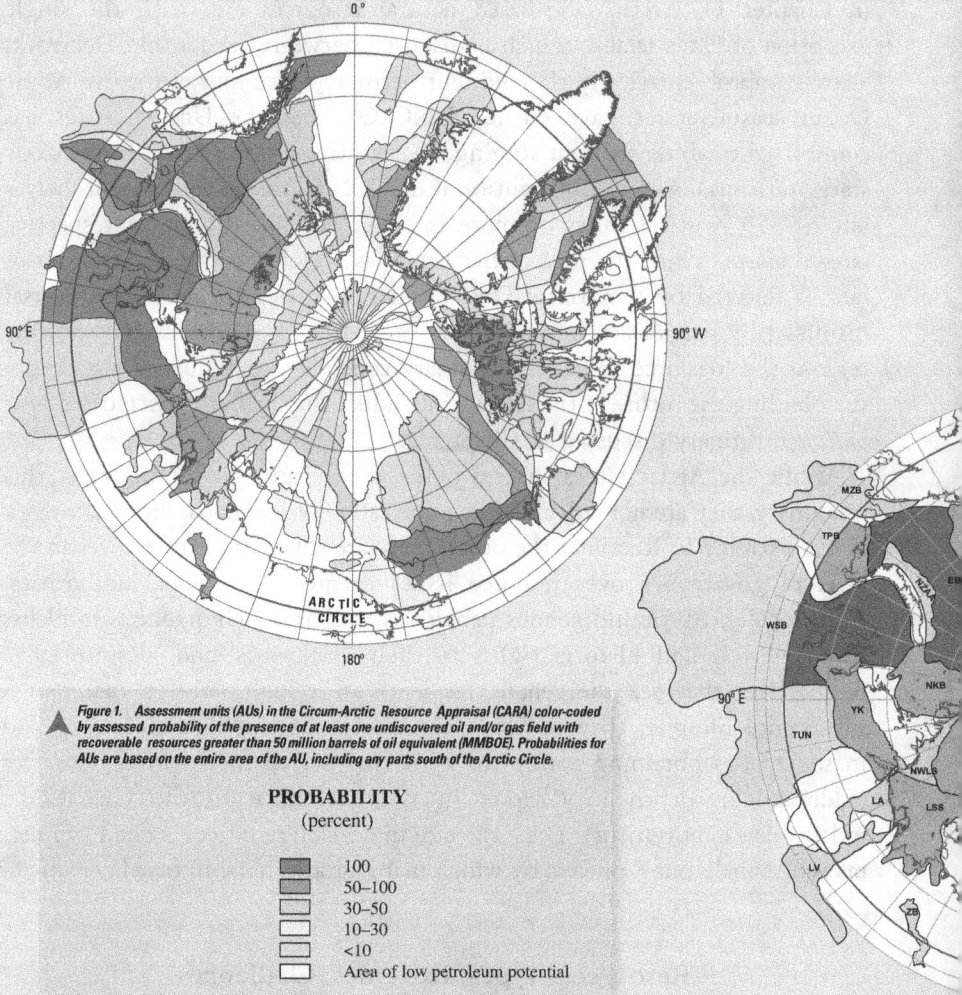

Figure 1.   Assessment units (AUs) in the Circum-Arctic Resource Appraisal (CARA) color-coded by assessed probability of the presence of at least one undiscovered oil and/or gas field with recoverable resources greater than 50 million barrels of oil equivalent (MMBOE). Probabilities for AUs are based on the entire area of the AU, including any parts south of the Arctic Circle.

PROBABILITY
(percent)

- 100
- 50–100
- 30–50
- 10–30
- <10
- Area of low petroleum potential

5.2   Petroleum Potential of Assessment Units and Provinces in the Circum-Arctic, United St Geological Survey.

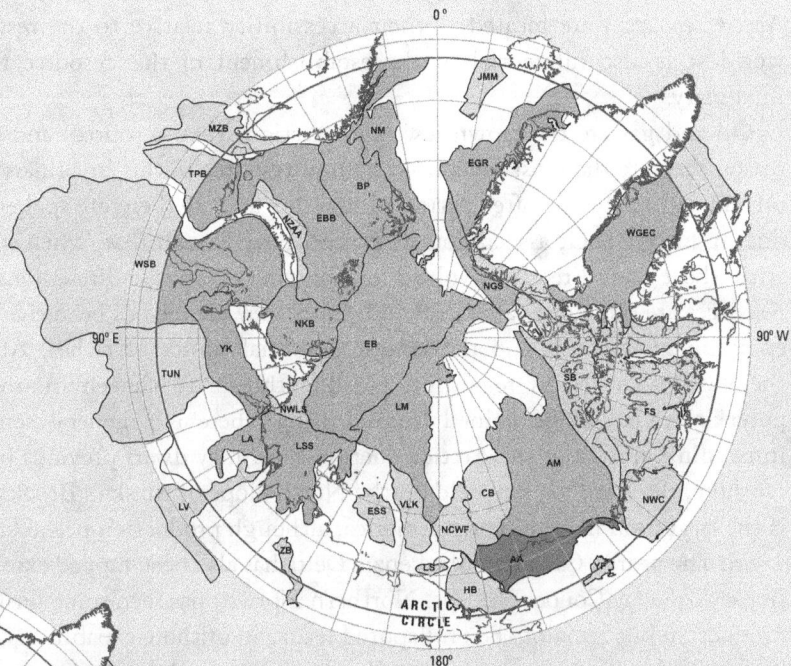

Figure 3. Provinces in the Circum-Arctic Resource Appraisal (CARA) color-coded for mean estimated undiscovered oil in oil fields. Only areas north of the Arctic Circle are included in the estimates. Province labels are the same as in table 1.

UNDISCOVERED OIL
(billion barrels)

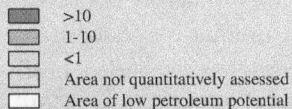

- >10
- 1-10
- <1
- Area not quantitatively assessed
- Area of low petroleum potential

Figure 2. Provinces in the Circum-Arctic Resource Appraisal (CARA) color-coded for mean estimated undiscovered gas. Only areas north of the Arctic Circle are included in the estimates. Province labels are the same as in table 1.

UNDISCOVERED GAS
(trillion cubic feet)

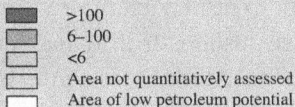

- >100
- 6–100
- <6
- Area not quantitatively assessed
- Area of low petroleum potential

Although the statistics on resources are impressive, and natural resource wealth can potentially bring lucrative returns on investments, *where* Arctic resources are located – their accessibility relative to the rest of the world – is also influential in the establishment of the 'resource frontier' imaginary.

Oil and gas are the crown jewels of Arctic natural resources and for this reason they are the most referenced natural resource in this imaginary. Crude oil and natural gas are often extracted together and are therefore spoken of as a dual industry. Their association with the Arctic is such that, when speaking with those not directly involved in the energy sector, discussions about resources are often only about oil and gas as they are seen as both the biggest environmental threat and the largest potential source of riches. Although there is no consensus regarding possible threats to the environment, or potential riches to be gained from drilling, there is a general sense that impending growth in the Arctic's energy sector may dwarf previous projects. Onshore, the Prudhoe Bay field on the North Slope of Alaska's Brooks Range is the largest oil field in North America, although production peaked there in 1988. The nearby Canadian Mackenzie Delta has also been tapped extensively, in particular by ConocoPhillips. Northern Norway has seen some onshore oil production but remains the undisputed leader in offshore capabilities. Russia has teamed with Norwegian Statoil to develop further its western Arctic onshore fields as well as offshore fields in the Barents Sea to complement its offshore operations in Pacific Russia.

Onshore oil and gas development is overshadowed by the potential of the Arctic's undersea fields. In 2001, Snøhvit, in northern Norway, became the first productive offshore gas project well above the Arctic Circle. Since then, Statoil has expanded north from its other North Atlantic fields into the Arctic. 'Statoil is certainly considered to be the world leader in operating in offshore Arctic environments; there is no question about that', was the upfront assessment offered by the director of a US oil and gas industry association. Russia has drilled offshore in the Arctic as well. After a 2011 deal with ExxonMobil to drill in the Kara Sea, Russian president Vladimir Putin said, 'It's scary to utter such figures' in reference to the expected value of the field. Extensive offshore drilling has yet to come to US Arctic waters. Despite leasing rights in Alaska's Beaufort and Chukchi Seas, in 2010 President Barack Obama extended a moratorium on further offshore drilling in the wake of the Deepwater Horizon disaster in the Gulf of Mexico. The moratorium was subsequently lifted but Shell, the leaseholder, has continued to face a number of setbacks. Although in many offshore areas the amounts of oil and gas are unknown, an executive of a drilling-related company referenced high expectations for Alaskan waters:

The potential is there for some more discoveries; otherwise people wouldn't be buying the leases and wouldn't be putting millions of dollars into planning and staging and getting everything ready to go. So they have to have some confidence.

In all of the littoral countries, state-owned, public–private and private oil and gas companies have taken advantage of improvements in drilling technology, sea-ice melt and a hungry market to leap into Arctic waters.

Though dominant economically, oil and gas are not the only valuable natural resources in the North. Mineral deposits are currently being mined or explored throughout the region. Silver, gold, copper, tungsten, zinc, coal, diamonds and uranium among other minerals are being exploited by multinational corporations from around the globe. Since mining in general is a more labor-intensive industry than oil, and because it has a history in the region dating back to the 1890s Yukon Gold Rush, this type of extraction has had a more sustained impact on local employment and economic development. A Canadian industry association representative was quick to note that new deposits of valuable minerals continue to be found, to the benefit of a broad section of northern residents: 'Since the discovery of diamonds in 1991, the GDP of the Northwest Territories has tripled, so that has brought employment to the aboriginal communities, to Northerners in general.' In Greenland, the potential for uranium mining is high (and returns are expected to be great), and thus in October 2013 Greenland's Prime Minister Aleqa Hammond straightforwardly explained,

We cannot live with unemployment and cost of living increases while our economy is at a standstill. It is therefore necessary that we eliminate zero tolerance towards uranium now.

A little more than a year earlier, in June 2012, Nunavut government officials had announced a policy that would allow uranium extraction so long as the uranium would be used for peaceful purposes and under the condition that its mining be conducted according to strict environmental and health guidelines.

Mining requires a much greater investment in infrastructure and typically reaps less profit than oil. As such, a mine is a more risky venture and the industry is itself tied to cheap sources of oil to power its operations. The upturn in oil prices has therefore slowed growth. 'It is typical where they are putting 200–300 million [dollars] in seismic work or mineral exploration and only one in ten or one in fifty come to commercial development' is how a Canadian official viewed the financial risk associated with minerals exploration. Mineral deposits in the North may be massive but mining there is an expensive venture. Thus, mining does not contribute as much to the 'gold rush' mentality as it did in the past.

Regarding fisheries, while commercial catches in the southern reaches of the Arctic Ocean have been pursued for centuries, they have intensified recently due to the opening of more summer waters and the subsequent movement of increased stocks into the North in combination with stricter international fishing regulations. Like mining, fishing also bridges the gap between local labor and global profits. While subsistence fishing was (and continues to be) practiced by indigenous communities, commercial fishing also has a long history in the Arctic. An Alaskan official observed that fishing is a matter of lifestyle for Alaska Natives, a major draw for summer visitors and a key economic resource for the state. Over-exploitation has led to some nations issuing moratoria in particular zones or for particular species, much to the chagrin of some coastal Arctic residents, who claim traditional rights to fish. A Norwegian official commented on the long tradition of fishing in his community, a way of life that is viewed not simply as a choice but as a right:

My community and all my neighboring communities are a thousand, two thousand year old fishing villages, and have had access to the free markets to export fish since at least the year 800, and there is a very strong feeling that this is the right of those communities.

Problematically, though, fish are a 'moving resource', so commercial ships regulated in one state's waters are able to continue their catch by moving into adjacent, less regulated or, in some instances, unpatrolled waters. This has led to cooperative agreements, for example between Norway and Russia, to avoid a 'tragedy of the commons', as both states control waters vital to fish stocks that straddle EEZ boundaries. However, moratoria and regulations can only be enforced within territorial and EEZ waters. This has led in particular to ships flagged in non-Arctic states heading to unregulated waters beyond state jurisdiction. This mode of fishing has been positively affected by climate change because commercial fishing can reach farther north and operate during more of the year as the Arctic Ocean's ice cover decreases. Although fisheries experts hold that, even with global warming and ice melting, large-scale commercial fisheries are unlikely to emerge in the Beaufort, Chukchi or Central Arctic Oceans because of low phytoplankton productivity and limited circulatory systems, the waters of the Kara Sea north of Russia, in particular, are likely to see a substantial increase in fish stocks. Other species, meanwhile, may confound existing agreements by migrating to new, more attractive grounds (viz. the ongoing conflict between Norway and Iceland brought about by changes in the migration patterns of mackerel stocks). A US Coast Guard official noted matter-of-factly how the drive for profits will push back the 'fisheries frontier':

> Well it [the Arctic Ocean] may not be covered in ice year-round anymore, and if it is commercially viable and the fish stocks and other species move north, and the fishing fleet sees a profit in it, then they will move north.

The Arctic is oil rich yet, perhaps counterintuitively, the prices for fuel in the North are steep. This fact slows resource development as most industries are dependent upon large and accessible sources of fuel under narrow seasonal time constraints. It may seem that this economic constraint, coupled with the harsh Arctic environment, would act to deter the harvesting of the commons but this is hardly the case. As a Norwegian oil industry representative explained, 'We realize that everything we do in the Arctic has to have a different standard because we need to handle more extreme temperatures, but it's not more difficult than actually extracting [oil elsewhere].' In fact, for many in the industry, the challenges of extraction, including remoteness, ice and extreme cold, are seen not so much as economic and material *problems* that hamper development, but rather as *opportunities* for overcoming something difficult. Whether it involves finding new technological solutions for frigid Arctic conditions or using brute force of labor to get to the resource, Arctic oil extraction perhaps resonates with the wild frontier experiences of the early Texas wildcatters. The fishing, mining and petroleum industries perpetuate in some circles a sentiment of 'struggle against the elements', and surviving and thriving in the Arctic contributes to each nation's sense of northern identity. This attitude is central to upholding the 'resource frontier' imaginary because extraction is not just about profits but includes the mythos of overcoming wild nature.

Filling the Arctic's 'Sovereignty Holes'

Paralleling this idealization of the Arctic as a frontier to be conquered is that of a region with 'sovereignty holes'. Despite some uncertainty about how far the continental shelves of the Arctic states extend into the Central Arctic Ocean, there is little debate regarding the framework under which Arctic Ocean resources are to be allocated. Although few of the 'holes' that are located in the Arctic beyond the 200 nautical mile limit of EEZs are likely to host commercial fisheries, and even fewer of the smaller 'holes' beyond the limits of outer continental shelves are likely to present commercially viable mineral extraction opportunities, the rhetoric of 'holes', where sovereign authority ceases to exist, is replete in the literature on northern resources. In particular, fishing industry officials and analysts cite the overfishing of pollock in the Bering Sea 'donut hole' during the 1970s and 1980s, when Japanese fishers – who claimed to be harvesting the Bering's international waters – were suspected of finding their catches within the US EEZ. As an Alaskan fisheries official recalled:

> And so what [the fisheries inspectors] did is they came into Anchorage here and they hired a small plane, a military plane, and they took one of those really old video cameras from that era, and they flew out there and found this whole Japanese fleet sitting in our zone, and as they began to take pictures [the Japanese] were putting canvasses over the [ships'] numbers.

While the problem of 'holes' may well be overstated, there *is* a general problem of enforcement in the areas of the region with productive fisheries, and this potentially could facilitate over-exploitation of fish species in the 'Arctic frontier'.

Perhaps the most significant 'holes' in the extension of territorial sovereignty to the Arctic are legal rather than geographic. Consider the example of the Chinese flag-hoisting and the opening of China's Arctic Scientific Research Station, as well as analogous initiatives by other Asian countries, European countries and the European Union: they have all become part of the Arctic discussion by establishing some sort of northern presence. An Arctic research executive claimed, 'There very literally is an opening of the Arctic. There is this sort of nationalistic interest in claiming territory or resource rights, and funding research has traditionally been a way to accomplish that: to establish a presence.' We can also see this intent from an EU official, who claimed, 'We should work together in order to ensure that we can exploit those resources, but at the same time ensure that we can protect them and conserve them for future generations.' Here, there is an explicit claim that 'we' (including the European Union) can exploit the resources, with attention given to their protection and conservation via science. Thus, a scientific presence in the Arctic can be used to support one nation's claims, or it can be used as a way to internationalize the North (and its resources).

In some instances, legal 'holes' are combined with geographic ones, as is the case for instance in the Norwegian territory of Svalbard where, under the terms of the Spitzbergen Treaty of 1920, any signatory may engage in economic or scientific activity. Indeed, this explains the popularity of Svalbard among non-Arctic countries that seek to establish an active presence – if not actually a claim to sovereign territory – in the Arctic through scientific research programs. Another method that has been used to establish a presence is via international shipping through the Northern Sea Route and Northwest Passage. In fact, shipping itself represents a 'hole' of sorts in the sovereign governance of Arctic activities, providing an important means for non-Arctic nations to claim a stake in the Arctic's future. Shipping provides an opening for non-Arctic states to establish themselves in the region for three reasons: because UNCLOS bestows considerable navigation rights to non-coastal states under which ships are flagged; because of opposition (from the United States in particular) to coastal state implementation of environmental regulations in Arctic waters that are seen as 'excessive' interpretations of UNCLOS Article 234; and because, unlike in many other regions, there is no specific shipping agreement for Arctic waters. In particular, Russia's efforts to regularize use of the Northern Sea Route by foreign nations' vessels, while reaffirming Russia's claim to sovereign jurisdiction, could also establish the presence of non-Arctic states (and companies) as Arctic 'players', which they could then use, for instance, to support requests for permanent observer status on the Arctic Council.

Finally, some non-Arctic states have sought to 'weasel in' to the Arctic (to use a term employed by one US official) by dealing directly with Arctic indigenous peoples or Arctic provinces. As noted in the previous chapter, Greenland is the most obvious example of a regional government undercutting national authority, in part through deals with outsiders. Beyond the example of Greenland, Alaska in the United States and the Canadian territories are granted various degrees of autonomy in their control of resources and have sought to attract commercial interest from around the world. As one Alaskan policy researcher put it, 'Alaska wants to sell stuff to the world.' In Canada, the Yukon and Northwest Territories have achieved 'devolved' status, which grants them an enhanced level of self-government. Even in Nunavut, however, where devolution has not occured, the territorial government is an important gatekeeper that can benefit by facilitating outside investors' access to the region. As a Canadian elected official noted, 'The Government of Nunavut is probably the most motivated and is well placed at the moment to gain benefits from resource development, because all the good lands have been selected by the Inuit.'

Some governments from outside the region have been able to use indigenous peoples' and northern provinces' relative autonomy to access resources. For example, in 2011 China became the third country (after Japan and South Korea) to sign an exclusive import pledge with Canada for seal meat, skins and oil. Although the seal trade deal was made with Canada, it granted access to a huge market for products that are traditionally and exclusively Inuit. A US State Department official suggested that non-Arctic governments, by working directly with key non-state Arctic stakeholders, sometimes are 'skipping the middle man' in the Arctic:

> They are having trade negotiations between the Inuit and the Chinese directly, not with the Canadian government, because the Inuit have mineral rights over some of their lands, so they are bypassing the Canadian government and going directly to indigenous folks, who are putting up on their website, 'We are open for business.'

This type of local trade deal illustrates that resource sovereignty in the Arctic can be flexible, and further that foreign powers have the ability to broker this flexible sovereignty by dealing with semi-autonomous regional governments. Thus, to gain a foothold in the Arctic, if a non-Arctic player cannot strike a deal with a national government, there may be opportunities to negotiate an agreement with a regional authority. The Arctic's 'sovereignty holes' may thereby offer a variety of imaginaries, where innovative cooperation among diverse political, corporate and social entities can yield collective benefits.

Practically, if the presence of outsiders is to be formally institutionalized, it seems likely that one mechanism for doing so will be through granting more non-Arctic states permanent observer status in the Arctic Council. Calls for expanding the Arctic Council have been strongly opposed by some, in particular the indigenous permanent participants, who fear that their power will be diluted (see Chapter 6). However, those who support expansion claim that not only should the region be globalized (and accordingly that the Arctic Council should be opened to include new permanent observers), but, more importantly, that the Arctic has always been globalized through centuries of trade and transit. For these individuals, the Arctic is and has been a global public good, a transport highway for international shipment and a multicultural Mediterranean of the North. It follows from this logic that the Arctic Council then could serve as an international institution that facilitates peaceful access to the North for Arctic and non-Arctic actors alike. If the Arctic Council could coordinate the northern activities of the global oil and gas industry, the global transportation industry and globalized extractive industries, then, the argument goes, the Arctic Council could become a forum for non-Arctic as well as Arctic nations who wish to guide its development.

Defending State Sovereignty

Many officials in Arctic states do see the Arctic as a frontier but they are quick to add that it is not a commons. To them, the Arctic is already spoken for and, notwithstanding the portions that lie beyond the limits of EEZs or outer continental shelf claims, the Arctic and its resources are already 'their backyard'. Some representatives from the Arctic states are perpetuating the idea that the region is indeed a traditional frontier, where natural resources are the prize in a race; they just see themselves as the participants and as the ultimate judges with regard to what will be won in the North. For example, a Russian active in one of the Arctic Council's working groups described the Arctic as an arena of 'corporate greediness' but went on to note that states, acting in coordination through the Arctic Council, could regulate this 'greediness'. He elaborated, stating that the Arctic Council was 'a bit more of a practical instrument in stabilization of this greediness or mess, because the level of participation by the states in this forum makes it possible to control emotions.'

In general, representatives of Arctic states justify their right to Arctic resources by citing the territorial powers granted them through UNCLOS (in their territorial waters, EEZs and outer continental shelves), coupled with the standard rights of sovereign states to control their land-based resource deposits. This vision is tied to nationalism and traditional state geopolitics, where an 'us versus them' realism often prevails. In this sense, some stakeholders see other states, or their federal governments, as obstacles to sustainable and locally beneficial resource development. Nationalistic sentiments were articulated by some respondents with regard to opening up permanent observer status in the Arctic Council to non-Arctic states, as they argued that, to the extent that collaborative governance is required in the Arctic region, involvement in defining this governance should be reserved for 'authentic' northern stakeholders with tenable, territorially defined rights and responsibilities.

So, while foreign nations may be transparent in their desire to access Arctic resources, according to representatives from the Arctic Five these resources are not up for grabs. A Danish official stated this directly:

Actually, foreign countries – they are welcome as visitors, and they are also welcome if they contribute somehow. They are welcome to send all their scientists, but . . . they are not there to reap the benefit of oil and gas.

In Canada, a similar sense of northern ownership can be found. Commenting on outsider interest in Arctic oil and gas, an industry representative offered, 'They've been staked in words but not law. So Canada has been adamant that this is our area, and so we expect to develop it, whenever conditions may be [appropriate].' Notwithstanding the uncertainties about precise boundaries that will persist until outer continental shelf claims have been submitted and assessed by the Commission on the Limits of the Continental Shelf, most participants feel that these issues will eventually be peacefully and easily resolved among Arctic states, opening the way for them to choose either to drill for oil and gas through national firms or to sell the drilling rights to corporate interests. As one US industry representative said with reference to the Beaufort Sea delimitation line, 'The only serious territorial dispute is between us [the United States] and Canada. And I don't think we're going to go to war over that. So this is kind of a non-issue too.'

Much of the 'race' mentality among the Arctic Five appears to emanate not from a sense that rights are undetermined but rather from the fear that failure to be first-to-market will bring about a competitive disadvantage. As a US Arctic development advocate declared, 'The Russians are drilling, the Norwegians are drilling, I mean, it's just us [who are not].' This position was taken further by a Canadian industry representative:

> The present government has advanced themselves on this issue versus the opposition. And it's partly in the context of recent [moves by] Norway and Russia and Denmark. There's tons of resources out there, there's a huge interest, and we better start taking development more seriously up there, because it's sovereignty.

The Canadian industry representative's reference to sovereignty suggests a clear connection between development of the Arctic resource frontier and notions of national identity, pride and state-building. Although this equating of resource extraction with national political and cultural goals is perhaps most evident in Canadian Prime Minister Stephen Harper's 'use it or lose it' declaration (i.e., use the resources or lose the territory), it is demonstrated by other countries of the region as well, whether resource extraction is connected with the identity and security of the nation (as in Canada, Russia and Norway), or with a sub-national, non-contiguous region (such as Greenland and Alaska). As was discussed in Chapter 4, in Greenland, oil and gas revenues are seen as the ticket to full independence from Denmark, and thus their exploitation is directly tied to the nationalist agenda. But a similar equation between resource extraction, development, sovereignty and wealth is made in more established northern nation-states as well. As one Canadian noted: 'We have half the continent, a tenth of the people and we should be much richer than we are. Why haven't we done better than we have?'

The Arctic occupies an important place in the Canadian psyche, as has been noted by a host of Canadian cultural commentators. In writing about 'the True North strong and free' (a line from the Canadian national anthem), Canadian sociologist Rob Shields has observed, 'southerners construe the North as a counter-balance to the civilized world of southern cities, yet the core of their own, personal Canadian identity.' A similar sense of the Arctic as the heart of the nation can be found in the Norwegian notion of 'Nordrområdende', which has been used to focus development and security initiatives on Norway's northernmost counties. The north, as expressed quite directly by a commentator in Norway, is vital to Norwegian identity: 'It's part of who we are, and I think we would probably like our spirit to believe that we are living in the Arctic even when we're not.' A spokesperson for a Norwegian environmental NGO, however, suggested that this type of northern nationalism is only a thinly veiled justification for more petroleum development:

> 'Nordområdende' becomes a catch-word, a priority, and there is a lot of rhetoric about bringing people together, and cooperation between the states. But if you look at the real priorities, what they are actually looking for is oil.

In Alaska, as in Greenland, the link between resource extraction, national identity and political security is more complex to characterize, given the cultural and geographic separation between individuals in the non-Arctic continental United States and Denmark and their respective Arctic resource frontiers of Alaska and Greenland. In addition to providing a basis for regional autonomy and identity movements (which, while more pronounced in Greenland, are also present in Alaska), residents of the non-Arctic regions have often found ways to incorporate frontier resource extraction into national identity and security narratives. For example, a Danish politician positioned Greenland as central to Denmark's national identity: 'And the Danes, we see ourselves as an Atlantic nation and Greenland and the Faeroes are a very vital part of our kingdom.' Many US respondents familiar with the position of the Arctic in the other Arctic Ocean littoral states concluded that the United States does not have a similar sense of what has been broadly referred to as 'Nordicity'. Americans in general do not see themselves as an Arctic country but as a country that has an Arctic resource frontier in Alaska. This vision of an intertwined union of 50 states clearly justifies a broad acceptance that Alaska's resources should be used for the benefit of the entire nation. Indeed, in the lower 48 states of the continental United States, investment in Alaskan oil and gas production is often promoted as a path toward national energy independence from foreign sources.

Complementing similar visions of national unity in the pursuit and possession of resources is the notion that competition within and among Arctic states will produce winners and losers in the Arctic resource race. For both Alaska and the Canadian Arctic territories, while devolution (to the various degrees that it has been implemented) has given regional governments more authority, they nonetheless feel pressure from their capitals to extract profits for the south. As one Canadian who had been active in implementing Canada's Arctic policy remarked, 'Most states go into the region and take [resources] out [to] somewhere else. It's a frontier for our resource sector.' Much of the intrastate conflict surrounds this very issue – the equitable division of benefits derived from resource access among northern residents and the rest of the nation. For many Alaskans, the act of piping what they see as 'their' Arctic resource wealth to Washington offers little reward. An Alaska Native activist offered this assessment: 'This is the crazy part, you know: We're the richest country, we are the richest natural resource state in the richest country in the world, and they have better health care in the villages in Cuba.' Within Arctic littoral states where regions are granted less autonomy over resources, residents have less control over resource management and receive fewer of the benefits that can come from resource development. In Russia, for example, resource-rich regions pay a hefty tax to the central government located in Moscow, which is where the majority of tax payments remain. In this manner, much of the revenue derived from Russia's resource wealth flows from the eastern regions to Moscow and St Petersburg. As a representative of the mining industry observed, 'Anybody who thinks that the Chukchi or others are going to have much to say about how Russian policies are defined and implemented just doesn't get it. This will be a Moscow-based set of decisions.'

Despite the differences in approaches to resource management, Arctic states' national policies throughout the region tend to coalesce around maximizing revenues from natural resource development. A representative from a Canadian indigenous organization characterized the attitude toward the North held by the bulk of southern Canadians as follows: 'I mean there's still this pot of gold at the end of the rainbow Klondike [myth]: "We are happy that the Arctic is ours because someday we [Canadians] are all going to get rich."'

Networks of Production

Government officials in Arctic states are not necessarily opposed to the 'corporatization' of the North, and representatives of industry are not opposed to resolution of Arctic international boundaries in so far as this can facilitate their operations. Despite the competition implied by the 'race for resources' metaphor, states and corporate entities can benefit alike from both private and public expansion of resource extraction; in fact, the boundaries between public, private, local and global blur in the Arctic. In a fundamental sense, if a state leases access to offshore oil and gas to a corporation, is the resource controlled by the leasing state, the corporation, or by the corporation's nation of registration? This ambiguity among interested parties is well illustrated by an industry association representative's comments on the global interest in Arctic gas: 'So many foreign companies are buying lands, producing the gas, or have plans to produce the gas and then ship it back to the terminal, where it can be sent to their home country.' Not only are corporations and states mutually dependent in their efforts to exploit the Arctic but, quite often, positions within corporate hierarchies and government bureaucracies are filled by people who regularly move between these two seemingly separate spheres. Further, there are regular exchanges of ideas (e.g., guidelines regarding best practices) and money (e.g., royalty payments and leases), in addition to personnel, between these groups, and with indigenous and environmental organizations as well.

Viewing the Arctic as a frontier commons and, by implication, the site of a free-for-all 'race for resources' is clearly an oversimplification. The dynamic interplay of people, funding, technology, research, governance and efforts to access resources is in fact better characterized as one of cooperation. If there is indeed a 'race' for access to the resource frontier, its participants share a desire to establish control over future profits from the North. Indeed, the participants' success depends very much on their mutual cooperation.

The expansion of control over Arctic resources relies heavily on groups of interconnected people working toward similar goals. But, as geographer Matthew Huber has noted, cooperation does not by necessity lead to positive outcomes: 'The story of oil is loaded with spectacular examples of how the state and capital combine to generate wealth through greed, war, and socio-environmental injustice.' In the Arctic, although both private corporations and governments may benefit from resource extraction, benefits are likely to accrue unevenly among all Arctic actors, producing a variety of losers as well as winners.

All the Arctic states earn revenue through the extraction of petrochemical resources, but precise national policies vary. An official from one of Canada's northern territories highlighted the differences between Canadian and American approaches to resource development:

From the development side, the Americans seem to have a different approach to doing business. For example, on the Alaska pipeline, the Americans are more willing to provide loan guarantees, but they are not too concerned with the fact that loan guarantees help so-called big oil. Whereas in Canada, our government, the federal government, is very leery, very cautious about not wanting to be seen that they are providing subsidies to oil companies that are making billions of dollars in profits.

Canada has to some extent devolved control of resources to the northern territories as well as the Inuit land claims organizations, thus adding additional levels of beneficiary. But global oil and gas as well as mining corporations are still motivated to drill and dig in Canada as they can reap great financial rewards for doing so. ConocoPhillips alone has 45 different significant discovery licenses in Canada. The Government of Greenland is quite straightforward in believing that extraction of nationalized resources will provide the final impetus for the territory's complete independence from Denmark. However, Greenlanders' lack of drilling expertise means that they will depend on production-sharing agreements to accomplish their resource facilitated independence. Soviet resource extraction was certainly focused on creating wealth for the state but even after the fall of that political regime Russia has relied on semi-privatized but largely state-owned entities such as Gazprom, the world's largest natural gas producer. Norway's Statoil is also a public–private entity, with two-thirds of the company controlled by the state. Statoil representatives directly acknowledge the ambiguous position of their company:

> Some people forget that we are a company and not a nation-state .... We are a Norwegian company, we are global, but Norway is our biggest base because this is where we generate most of our revenues, so we have naturally slightly more focus on Norway.

Natural resource revenues in the Arctic are 'owned' by varying combinations of public and private entities. Because of this, sovereignty over resources rests on a complex mixture of territorial control and legal institutions that grant sovereign rights.

If there were separate, individual entities in pursuit of Arctic resources, then these likely indeed would be competitive, as each would pursue its own interest by trying to prevent other players from gaining an advantage. While elements of this type of gamesmanship are at times visible, nonetheless the Arctic is better characterized by what might be called *dynamic cooperation*. This is not to say that environmental NGOs are not often at odds with resource developers, or that indigenous groups find certain initiatives by environmental NGOs patronizing or contrary to their lifeways. But more broadly, because of the physical constraints of production in Arctic climates, the remoteness of sites and the relative lack of labor, the development of resources, or equally the resistance to development, people are compelled to work together. This cooperation suggests that a network of relations helps to determine which lands and waters are accessed and how profits are split, rather than the more popular vision that there is a race and that one nation might somehow win the Arctic 'prize'.

In developing the Arctic's oil, cooperation is clearly indicated in the manner that rights to drill in many of the fields are shared by two or more companies. For example, rights to exploit the Amauligak field in the Canadian Arctic are 54 percent owned by ConocoPhillips, but with minority shares held by Chevron (36 percent) and ExxonMobil (7.5 percent) as well as others (the remaining 2.5 percent). In Russia, ExxonMobil holds a 33 percent share in the venture established to drill the West Siberian fields but the controlling 66 percent stake is held by Rosneft, the largest Russian state-private oil company. Reflecting on his previous experiences in a more competitive industry, a Statoil executive noted, 'Here I find that yes, of course, we are competitors, but they can be our partners the next day.'

A similar sharing of access to deposits occurs in Arctic mining. Alaska's Pebble copper mine is owned and operated by several multinational corporations based in the United States, Canada and the United Kingdom. Arctic fishing also involves significant cooperation. Since fish are a mobile resource, and many stocks have lengthy migrations that do not respect national borders, cooperative management has become the norm. This can be seen in the management of pollock stocks between the United States and Russia and similarly between Norway and Russia. Fishing experts tend to agree that this cooperation is the only way to ensure sustained extraction from Arctic fisheries, as a competitive approach to fisheries management would lead to the complete crash of a resource that is already threatened with over-exploitation. Perhaps the most dramatic cooperation is occurring in outer continental shelf mapping, discussed in Chapter 2. The very purpose of this activity is to enable each Arctic state to make an exclusive claim to mineral rights over a finite quantity of Arctic seabed. And yet even here Canada is cooperating with the United States and Denmark in its western and eastern boundary waters respectively, notwithstanding its ongoing disputes over the Beaufort Sea and Hans Island.

A norm of cooperation in the resource-extraction sector extends as well to day-to-day operations, as companies and governments work together to ensure the continued flow of profits for each. Several respondents in the oil and gas industry noted that companies share equipment, particularly helicopters, which are often the only means of transportation in the infrastructure-poor Arctic. Due to lack of infrastructure, safety is also a domain where cooperation is standard. One respondent explained how, if someone is injured in an Alaska Native village and a mining company plane is the only way to get that individual to a hospital, the company will work together with that community, even if residents of the community wholeheartedly oppose the company's presence in the North. This type of cooperation is mirrored in the Arctic Council's first binding agreement on search and rescue, signed in 2011, and its 2013 agreement on oil spill prevention and remediation. Finally, cooperation is notable between government regulators and industry representatives. Although elsewhere industries are often resistant to government regulation of their activities, there is an understanding, especially in the oil and gas industry, that, in the Arctic, regulations may be needed. Industry representatives note that an unregulated, offshore oil project would never be insured for loss and that in this sense they need regulation. For instance, an official from Statoil noted that the company's reputation in environmental and workplace safety was due to a lengthy vetting process of projects, substantial environmental regulations imposed by the Norwegian government and, overall, a careful and steady approach to resource development.

Of course, not all northern peoples approve of this shared drive to exploit. As a representative from an indigenous people's organization in Alaska explained, the revolving door between industry and regulators is not necessarily an appropriate cooperative arrangement for resolving environmental issues. Addressing the appointment of a former leader of the fishing industry to supervise state environmental fisheries management, this representative asked, 'Does he want to resolve issues that are associated with the fisheries?... Hell no, he's interested in figuring out how to catch more fish.' To the disappointment of indigenous and environmental groups, cooperation between state and corporate entities drives resource exploitation, producing an uneven distribution of opportunities and profits, as well as losses, for Arctic stakeholders.

A Race to Cooperate

With regard to resources, sovereignty is less straightforward than it may seem. The vision of a race across the Arctic resource frontier derives from questions of resource ownership and from the imaginary of a cultureless Arctic, devoid of rules and regulations, that is simply available for 'pioneers' from outside the region, or that is already being claimed by well-positioned insiders. Often this is the imaginary described by observers from beyond the Arctic, whose views are tainted by outdated (and themselves idealized) visions of the Yukon Gold Rush or that are influenced by satellite photos of record-breaking summer ice melts. According to this 'resource frontier' imaginary, the Arctic is an arena of opportunity, whose resources are available for the taking by solitary individual, corporate or state entrepreneurs.

The North's vast resource wealth indeed remains largely unexplored. In terms of oil and gas, offshore deposits may hold the greatest of the Arctic's wealth, but onshore fields are large, profitable and potentially will provide suitable infrastructure for transporting offshore oil. Mining has been prosperous in the North for more than a century, and with a larger variety of metals now in global demand this industry will continue to diversify and expand in the Arctic. Commercial fishing holds a privileged position in the heritage of the Arctic and, despite some dwindling stocks, the opening of waters during the polar summer has made areas of the Arctic potentially even more valuable for international fishing fleets.

Like the *terra nullius* imaginary discussed in Chapter 2, few actors involved in the Arctic see the region as a global frontier commons of unregulated prospects from which all are to gain from resource-extraction opportunities. Certainly this vision has been dismissed by the Arctic Five, who stress that resources in the bulk of the region are regulated by individual states or, in the case of the Central Arctic Ocean, the state community's regime for the high seas (UNCLOS). However, claims that the 'race' is over, or that it never has existed, are made somewhat ambiguous by non-Arctic states' participation in the realms of Arctic science and research, transportation, state–corporate extraction and the push for a more inclusive Arctic Council. Multinational corporations further complicate traditional assumptions of state sovereignty as they extract tremendous profits, often through leases to resources that are controlled by the states themselves. Sovereignty over Arctic resources is thus somewhat flexible, and can be thought of as the right to a percentage of profits rather than outright control over territory.

Many Arctic commentators perpetuate the vision of a resource frontier and commons, and this has facilitated the rhetoric of a 'race' for resources. While there may indeed be a 'race' as resource-extraction and transport opportunities open up, the race is fundamentally cooperative rather than competitive, as the various state and non-state actors work together to lay out the track on which they can all run. The end goal, for the major players, and one that is sanctioned by *both* the modern state system and the imaginary of the Arctic as a resource frontier, is to realize value and enhance national security (in the broadest sense of the word) by moving material and profit from the higher latitudes to the more southerly ones.

At the same time, however, while cooperation establishes networks of resource extractive interests, the balance of influence is unequal. There are important players, including indigenous peoples and environmental NGOs, who are not empowered in the same ways as states and global oil companies. It is to these groups and *their* Arctic imaginaries that we turn in Chapters 6 and 7.

6.1 General Assembly of the Inuit Circumpolar Council, Nuuk, Greenland.

CHAPTER 6

TRANSCENDENT NATIONHOOD

The Inuit are among the most populous of the indigenous people who live in the circumpolar Arctic, spanning four countries, from Russia's eastern edge, across to Alaska, Canada and Greenland. The various Inuit groups, who today amount to some 150,000 people, have resided in the far North for millennia, sharing common cultural attributes and languages. The Inuit Circumpolar Council (ICC) is one of the largest and most influential international non-governmental organizations representing the interests of the native Inuit inhabitants and serves as one of the six permanent participants representing Arctic indigenous peoples on the Arctic Council. In the summer of 2010, the Inuit Circumpolar Council met for its quadrennial General Assembly in Nuuk, Greenland, and the stakes could not have been higher.

Change was gripping the Arctic: with the warming climate the High North was appearing to many as a new frontier of economic opportunity and wealth. Corporations and national governments were clamoring to position themselves favorably to get a piece of something that nonetheless was still quite intangible and of questionable worth. As the media and the southern populations increasingly looked northward, the Inuit found themselves confronting the largest wave yet of interest, power and money, breaking over their native lands and waters. The Inuit, more than ever, were feeling the need to make themselves heard.[1]

Of course, they had been working diligently to do just that for the last four decades, but now, with so many newcomers on the scene, they had to raise their voice even more than before. As one Inuit leader interviewed explained, 'Unless the Inuit can punch above their weight, they're going to be totally irrelevant. So you have to be organized, to be conscious over-achievers.' With the challenge of over-achieving looming over them, the General Assembly in Nuuk, nonetheless, proceeded not unlike those of the past: an introductory ceremony was conducted with traditional singing and dancing. Many of the delegates knew each other, there was much embracing and backslapping.

The leadership consisted of familiar faces, faces that bore the marks of experience in the prolonged fight for the indigenous rights of the Inuit. As the assembly progressed there were many talks given by invited guests, politicians, academics and activists that spelled out the challenges facing the Arctic. Some ICC members spoke as well, and again it was the established old guard, the battle-tested generation that had been with the ICC since its very beginnings. These leaders spoke of the past, of the old ways, of a culture in danger and the need to stand up to the encroachment of the outside world and its values and rationales.

The traditional values on display at the ICC meeting were especially enhanced by the location in Nuuk, which meant that many of the Greenlandic self-rule government officials were present. As was discussed in Chapter 4, Greenland, although primarily Inuit, has been moving in quite the opposite direction of the ICC. In contrast to the older generation of the ICC, the reigning government of Greenland, itself newly in power, consists of a younger generation, educated in Denmark and much more comfortable with Western norms and lifestyles. Greenland, though a moral and financial supporter of the ICC, has significantly parted ways with the ICC in its political strategies and aims, being much more accommodating to the basic Western model of capitalism and sovereignty. With this new Greenlandic direction, the ICC's General Assembly in Nuuk made for an interesting crossing of paths that brought many questions to the fore. What was the future of the ICC vision? Was this simply a Sisyphusian battle against the inevitable incorporation of the Inuit people into the established political and economic systems of the various nation-states in which they resided?

In short, the answer is 'no', or at least 'probably not'. But to understand this answer requires a much closer look at the ICC, its past work and its current aims and struggles. It also requires a sustained engagement with the political potential of a national identity that spans not just the boundaries of sovereign states but also, as the ICC proudly asserts, the divides between land, water and ice that are used within the modern state system to divide the world.

Constructing an Inuit Identity

It is somewhat ironic that it was a non-Inuit Dane, Knud Rasmussen, who in many ways initiated the recognition of a transnational Inuit identity, an identity that the Inuit have honed and projected with impressive political efficacy. Rasmussen, the son of a Danish missionary father and a part-Inuit mother who grew up and lived in Greenland, was an explorer and anthropologist with extremely close ties to the Inuit of Greenland. Fluent in the Inuit language of

Inuktitut and having acquired numerous Inuit skills, such as dog-sledging, Rasmussen made it his mission to gain a greater understanding of the Inuit people, not only within Greenland but across the Arctic. In 1921, Rasmussen, together with two invaluable guides from the Thule region of Greenland, embarked on his last and most impressive Fifth Thule Expedition. This expedition aimed to cross the entire region of Inuit habitation by dog sledge. By conducting this journey Rasmussen's goal was not simply to prove the possibility of dog-sledge travel across such vast tracts of territory: he also hoped to demonstrate the strong cultural ties that joined the Inuit people, in order to create a better understanding of these people as one interlinked group.

In addition to collecting stories and amulets from the Inuit, Rasmussen also spoke to them about the Inuit he had met on the journey, thus reaffirming what the Inuit had known but had never fully verified: that an extensive Inuit culture stretched from Greenland across the North American Arctic. When Rasmussen arrived at the western coast of Alaska his intention was to cross the Bering Strait, to confirm and document the existence of a small group of Inuit on the Asian continent, in the Arctic territory of Chukotka. The only thing holding Rasmussen back were the visa papers from Russia, which were taking a suspiciously long time to arrive. As weeks passed with no word, Rasmussen lost patience and embarked clandestinely across the Bering Strait. On this last leg of his journey Rasmussen did indeed find the Inuit of Chukotka and told them of his travels and his intentions of documenting Inuit culture. On the first evening of his formal interviews, Russian customs officials came knocking at the door and quickly proceeded to deport the rogue Danish explorer. Upon his return to Alaska a telegraph was awaiting him – the visa from Russia had arrived but it was too late: after four long years it was time to return home to Greenland.

Despite the setback in Russia, it can be argued that Rasmussen achieved for the Inuit a critical task for their future organization. He had identified and publicized an interconnected, albeit quite isolated Inuit culture that spanned thousands of miles across the Arctic and four sovereign states. With this newly awakened imaginary of a geographically expansive, shared culture, the Inuit had the initial inspiration to become empowered. They were not completely alone, despite their small communities and the great distances between them. They were, in fact, one people, over 100,000 strong, who not only shared a common way of being in the world but also common interests. This realization, although it took a few decades to cultivate, was becoming ever more critical as southern populations were steadily creeping northward, with ever greater impact.

Although specific histories of exploration and immigration differed across the region, significant contact between the indigenous peoples of the

North and the European descendant populations farther south began in the mid-nineteenth century, initiated primarily by a search to profit from the renewable resources of the Arctic region. By land, settlers came to trap and set up trading stations. By sea, whalers came to hunt the migrating whales for their precious blubber. In the beginning, these activities coincided with the Inuit's subsistence lifestyle without great interference. However, over time, the renewable resources began to dwindle to the point that some species, such as the prized Bow Whale, were hunted almost to extinction. The vanishing of the whale stocks, which were so critical to the livelihoods of the predominately coastal Inuit, was a first clear warning that the commercial interests of the south were not in tune with the subsistence cultures existing in the Arctic.

Eventually the Industrial Revolution and its need for resources and energy also turned northward. There were vast industrial mineral deposits to exploit, hydroelectric dams to be built and, eventually, oil and gas to be extracted. Oil was found in Prudhoe Bay, Alaska in the 1960s, the largest oil find in the United States to date. Shortly thereafter a veritable oil rush occurred in the Arctic North, spanning from Alaska to Greenland. With US peak oil production estimated to have occurred in the mid-1970s, the large finds in Alaska and then Canada were like a godsend to the oil-guzzling US economy. For the Inuit this development meant an entirely new chapter in their relationship with the south. Not only was there a steady stream of non-Inuit into the Arctic region to prospect and extract oil and gas but the impact on subsistence livelihoods in terms of impending offshore drilling, road building, increased air traffic and the construction and maintenance of pipelines was becoming an ever-present concern.

With this turn northward, governments, with their seats of power far to the south, also began for the first time to consider systematic policies to be applied to the scattered and relatively small indigenous populations that lived there. Beginning in the 1950s the impulse on the part of Canada and Denmark was essentially the same: namely, to incorporate these populations into the dominant societies of the south. This essentially meant two things: ending the Inuit's nomadic tendencies by moving them into permanent settlements, and sending the children of these settlements to a number of specifically established schools where they could receive a standard 'Western' education. Due to the remote and scattered nature of the Inuit settlements, this latter policy generally meant that Inuit children, particularly for high school, had to leave their settlements and be boarded at an assigned school, which most often meant a move to the regional capital.

It is not hard to imagine the disruptive effect that these policies had on Inuit society and culture. The settlements, often located far from the traditional hunting grounds, became dependent on goods and services subsidized by the state. Due to the lack of economic activities in these settlements, and hence high unemployment, the Inuit were given government welfare payments to make ends meet. The values and knowledge of the Inuit, removed from their traditional means of subsistence, had little meaning or worth in these settlements. With the task of education now assigned to the state, the role of the elders diminished. Children returned from their far-off schools speaking a different language, with little knowledge of, and diminishing interest in, the old ways. Christianity became the dominant form of spirituality. As a consequence of all these policies of assimilation, the Inuit were steadily being transformed from a resilient, independent and self-sustaining people, albeit one that constantly had to contend with hunger, to one that was increasingly dependent on external government support. The fabric of Inuit society was subsequently coming undone, and with the fateful introduction of alcohol, the final result was widespread dysfunctionality, witnessed in extremely high levels of poverty, domestic violence, sexual abuse and suicide.

Their lives upended and their identities remade as citizens of distant nations, some Inuit livelihoods were further manipulated to serve the goals of southern capitals. Thus, while most Inuit were concentrated in regional centers to facilitate assimilation, others were sent in the other direction. In Canada, in 1953, eight Inuit families were relocated from northern Québec to Grise Fiord, more than 1,000 miles away on Canada's northernmost island, Ellesmere Island. The families were relocated to bolster Canada's claim that Ellesmere, which up to that point had been uninhabited, was part of Canada's sovereign territory. Even today, Canadian government officials routinely cite the presence of (Inuit) Canadian citizens along its northern frontier to bolster Canadian sovereignty claims (see Chapter 3). Whether assimilated through relocation to regional towns or through forced settlement at far-flung outposts, there is a long history of nation-states imposing national identities onto Inuit individuals in order to support state-building projects.

Today, the impacts that southern non-Inuit cultures have had on the Inuit has become an integral part of the story they tell of themselves. It is always lingering in the background. For instance, upon asking an Inuit Greenlander about the possible cultural cost of impending Greenlandic industrialization, he quickly responded, 'A cultural cost? I think the cultural cost, uh, started in 1721', which was the date of the first Christian settlement in Nuuk founded by Hans Egede, who, incidentally, still stands proudly in statue form overlooking the city that he had once christened Godthåb. The non-Inuit who have a relation to the North are also acutely aware of the story. When asked about the differences that the Inuit of Greenland have with the Danish government, one Danish official replied, 'I think their anger is not really about the money. The anger is that they have been victims of what they call cultural imperialism.'

Confronted with the rapid changes that were being imposed throughout the Arctic region and on its peoples, many Inuit leaders responded in a way that southern populations, in their ignorance, or perhaps their hubris, had barely anticipated. The Inuit demanded to know what gave the southern populations the right to do as they pleased in the North without asking for permission or even input from the native populations who had lived in and used this region for millennia. The Inuit people, upon being informed that they were, in fact, citizens of a sovereign state that maintained ultimate control of the lands that they inhabited, responded by stating that they, actually, had never been asked whether they wanted to be such citizens, nor had they, as opposed to other native groups, ever signed a treaty or agreement that entailed a surrender of land.

These conversations were starting to happen in the 1960s, a time when the United Nations Declaration on Human Rights had been in existence for more than a decade and when the emerging 'West' was presenting itself as the protector of individual liberties in the face of communist ideologies that, it was said, would squash people's ability to attain self-determination and freedom. Forcing indigenous people *en masse* off their land through coercion, war or trickery was no longer an acceptable prospect. The assimilation process was also running into ever greater resistance as the youth, who were now graduating with western educations, failed to voluntarily give up their culture. Instead the Inuit had new leaders who now understood the legal, social and political intricacies of the dominant southern society. Some of these leaders were claiming that a cultural genocide was underway. And, as the land issue began to make its way through legal analyses and court hearings, it became increasingly apparent that the so-called sovereign states indeed had 'a problem' with the Inuit. Both sides felt pressure to come to a deal. Inuit leaders, particularly in Alaska and Canada, insisted that they would not agree to resource-extraction activities, or the construction of pipelines, unless the land issue was settled. Knowing that corporations were unwilling to risk investments in a hostile environment, the US and Canadian governments had an incentive to reach agreements with the Inuit leadership. The Inuit leadership, in turn, feared that if they remained too intransigent, governments would circumvent them, leaving them with nothing.

Alaska was the first to come to an agreement. In 1971, the US government signed the Alaska Native Claims Settlement Act with the indigenous peoples of Alaska, including the Inuit. In this treaty it was stipulated that Alaska Natives would receive full ownership of their traditional homelands. For the rest of the land they would be given monetary compensation and exclusive hunting rights, as well as increased self-governance. The money would be used to create for-profit corporations of which currently living Alaska Natives would be made shareholders. As for the subsoil, the Alaska Natives would receive a 2 percent royalty payment on all mineral development.

Denmark and Canada quickly followed suit. In 1979 Greenland obtained 'home rule' status, which transferred considerable authority to the Greenlandic parliament, while also guaranteeing specific land rights to the Greenlanders. Canada, in turn, followed the Alaskan example more closely. The first land claims agreement was implemented in northern Québec in 1975 to make way for the expansive James Bay hydroelectric project, which was to displace several Inuit and Cree communities. The largest of the Canadian land claims agreements, however, was signed in 1993 and resulted in the creation of Nunavut, a massive self-governing territory that incorporates what had been the northern and eastern portions of the Northwest Territories. Since that time, almost all the Inuit land in Canada has come under some form of land claims settlement. As in the first model arrangement in Alaska, the Canadian agreements all entail some form of land-ownership transfer to the Inuit, certain land-use rights in the territory that has been surrendered, self-governance arrangements and royalties from extractive industries.

The settlement of land claims was a significant breakthrough for the furthering of Inuit empowerment, particularly in Canada. As one representative of a Canadian Inuit corporation stated,

> Now we have Nunavut, now we have the land claims agreements individually, there isn't one Inuk in the northern regions in Canada that isn't represented by land claims, so land claims agreements ... have paved the way for us to believe that we should be able to have self-determination.

Yet one aspect of the various agreements with the Inuit that was non-negotiable, and to which the Inuit generally acceded at the time (with Greenland being a partial exception), was that the federal government (i.e., the state) was to retain full authority over issues of sovereignty, defense and foreign relations, thereby nipping in the bud any inclinations on the part of the Inuit to call for outright independence.

Creating the ICC

The Inuit would in many fundamental ways remain captive to the state in which they resided, and, importantly, their relationship to the state would remain their primary political starting point. Yet, the Inuit were a transnational people, spanning four sovereign states. It is within this context that the Inuit leadership decided to launch the transnational Inuit Circumpolar Conference (later renamed the Inuit Circumpolar Council) to pursue the interests of the Inuit people across the entire Arctic region. The first general assembly was held in 1977, in Barrow, Alaska, just over 50 years after Knud Rasmussen's visit there during his Fifth Thule Expedition. Now, the Inuit were seeking to unite along the cultural lines that Knud Rasmussen had worked to trace decades earlier, and the primary aim was to achieve the much contested goal of self-determination.

The first chair of the ICC was Eben Hopson, considered to be the founding father of the organization. Hopson was also the mayor of Alaska's North Slope Borough (of which Barrow is the capital) at that time. It was his vision to utilize the ICC as an organization and a forum in which the Inuit could share their regional experiences, come together as united people and advocate Inuit interests on an international scale. Over the past three decades the ICC has had remarkable success in pursuing Hopson's dream. To this day, the declarations emanating from the general assembly of the ICC, which is convened every four years, pay their respect to Eben Hopson, who had the vision and the leadership to bring the ICC to fruition.

The importance of the ICC for the Inuit people cannot be overstated. Given the limited powers attributed to the Inuit with regard to international affairs in the various land claim agreements made with their respective governments, the ICC has offered the Inuit an organization with which they can bring their collective international voice to bear on domestic issues, as well as providing the Inuit with a critical mouthpiece on issues that transcend national policy making. The cornerstone of the organization, as envisioned by Hopson, was to be the cultural integrity of the Inuit people understood as a nation, but one that is not in pursuit of a nation-state. Already in 1975, Hopson declared:

> We Eskimo people of Alaska, Canada, Greenland, and eventually the Soviet Union, can join together to meet common problems posed by industrial society encroaching upon our land, our communities, and our traditions. We ... are an international community sharing common language, culture, and a common land along the Arctic coast of Siberia, Alaska, Canada and Greenland. Although not a nation-state, as a people we do constitute a nation. This is important not because nationalism solves our problems but because our common nationality is the basis of our present attempt to find solutions to our common age old problem of survival.

Yet, nations, as scholars increasingly point out, are essentially imagined communities with constructed histories, and thus are practically never naturally occurring; they are created and maintained. Hence, initially sparked by the travels of Rasmussen, the ICC became the body tasked with fostering the national identity of the Inuit people, and as such it has taken on a role that transcends that of a traditional NGO by positioning itself as a political representative of the Inuit people on the international scene.

In the early years of the ICC, the job of creating a sense of coherence and shared culture was thus of primary importance. The Inuit language, spoken in various dialects across Inuit society, had fallen into disuse in many communities. There were few connections between the communities in terms of communication or transport infrastructure. The respective states to which these Inuit belonged had left their distinctive cultural marks. Lastly, the Inuit of Chukotka were still largely unreached, hidden behind the Iron Curtain of the Soviet Union.

To deal with this situation, the ICC moved on several fronts. To begin, the ICC looked to the United Nations for discursive and legal support. In order of their emergence, the ICC has made reference to the texts of the International Covenant on Economic, Social and Cultural Rights (1976), the International Labor Organization's Indigenous and Tribal People's Convention (1989), and the Declaration on the Rights of Indigenous Peoples (2007). In accessing this international rights discourse, the ICC has moved to protect and foster the Inuit language and culture, by, for example, successfully challenging state assimilation policies. One Greenlandic parliamentarian we interviewed explained her life growing up with an Inuit mother and a Danish father. At the time of her childhood she was denied the opportunity to attend an Inuit school, where they were teaching in the local Inuktitut language, which she spoke fluently. When the ICC was established in Greenland, however, there was a concerted push to allow all Greenlandic children the opportunity to learn in their local language. As she stated:

> In the early 1980s, when the ICC came ... they would help us to be a Greenlandic person .... They managed to say to me, 'You are a Greenlandic citizen and you're not a Dane,' in my mind. I am very grateful that the ICC has been there for so many years because it helped us a lot in the culture but also with our language.

The ICC also instituted quadrennial General Assemblies. These were not only intended to provide a forum where the Inuit leadership from around the Arctic could meet to discuss their common political ground; they have also served as cultural events in which the various Inuit people share their culture in the form of dance, music, stories, crafts and food. The ICC General Assemblies had become a veritable cultural spectacle that is made available to all Inuit in the hosting town.

Efforts were also made to reach out and include the Inuit of Chukotka who were not allowed by the Soviet government to attend the first General Assemblies. Working with the USSR to assure their presence in the ICC, one remarkable confidence-building step entailed a journey by a group of Alaskan Inupiats who embarked on a trip to visit their close relatives across the Bering Strait in Chukotka. Despite being only a few hundred miles apart, the Alaskans were forced to travel around the entire globe to reach their kin because there were no transportation routes connecting them. Shortly after this visit, the representatives of the Inuit of Chukotka were allowed to travel to the General Assemblies.

To this day the ICC continues to maintain and foster a coherent and unified front, rooted in the common story of the Inuit and their struggle to survive and thrive in the Arctic region. Yet many challenges remain, as stark differences between Inuit groups persist. For one, Greenland is geographically unique. The world's largest island exists far removed from the country that claims it, Denmark. This is perhaps one of the reasons that the Inuit of Greenland, who make up the large majority of residents there, are less inclined to identify as citizens of the state in which they reside. Given this situation, Greenland is the only Inuit territory in which there is serious consideration of becoming a fully independent and sovereign Inuit state (see Chapter 4). In Canada, by contrast, the North is not only geographically contiguous: it is often heralded as the nation's heartland, an appellation that, while not necessarily bestowed with the Inuit in mind, has given the indigenous peoples of the region a certain status as the embodiment of the national ideal. The Inuit of the United States and Russia are arguably somewhere in the middle, as they reside within frontier regions (Alaska and Chukotka, respectively) that, although marginal, are still considered part of the integral territory of the nation-state, and with majority settler populations. For its part, Chukotka, though now part of a somewhat more open Russia, is still isolated, and the rights of the Inuit there are much less formalized in comparison with Greenland, Canada and Alaska.

Greenland is also considered more developed and better off than most of the Inuit territories in North America and Russia, although the percentage of Inuit dependent on traditional subsistence hunting and fishing in Greenland is also much less. Yet the greater degree of urbanization in Greenland has also led to the Greenlandic Inuit acquiring the highest level of resources. As the Greenlandic parliamentarian quoted above proudly stated,

> I have heard that in some places in Canada the people don't have university degrees. Here we have our own university. We have our own teacher training school. Yes, we have a lot of schools that allow people to remain Greenlandic.

Due to the infrastructure that directly caters to the Inuit population, the Greenlandic language is also much better preserved. One Alaska Native activist told us about his father's impressions of Greenland, particularly regarding the language:

> He thought it was cool, the language was the language ... the radio station when it goes on is in the language, the street signs are in the language, the newspaper is in the language. You could go to the store and speak in the language, and that doesn't happen here, that doesn't happen in Barrow anymore, not typically. If that happens it's almost because people are making it happen; it is not the natural pulse.

Yet there also seems to be ambivalence about this difference. An official in the Danish government, with extensive experience working with the Inuit, pointed to this difference in the following way:

> It's very interesting because I've been very good friends with the former premier of Nunavut, Paul Okalik, and whenever he came to Greenland, you know, he was always very critical. He kept saying, 'Oh, you're so European,' in a negative sort of way. And you know, he was so envious. I mean he felt that this was so great. On the other hand, people of Greenland look to the Inuit of Canada and Alaska to find sort of the real Inuk spirit. They still have, you know, the Drum Dance and the *amauts*, you know, the bag to carry your kids and stuff like that. So there's a ping pong going on there.

This distinction between the modern Greenlandic experience and the more traditional North American one was also brought up in an interview with a key Greenlandic government official:

> Now in Alaska, I can see, they are living the cultural life. I love it, because it is as if that half of my person is being freed. When I am returning, the intellectual, the Westernized part of me, the Danish

part, I think is being freed. So I have a constant feeling of being a half human, half person.

With these differences there are also differing views and suggestions regarding policy. For its part, the ICC has striven to navigate these differences by focusing on the commonalities and the overlap of interests. As a Canadian Inuit working for one of the large Inuit corporations stated to us,

There have to be lines of tension and different approaches. But at the same time you want to tell the world the ICC represents a single people that live in four countries and they are quite capable of coordinating their action when their collective interests are at stake. And I think that's been a really impressive international story.

Or, as stated by an up-and-coming young official in the Greenlandic government:

We might have some competing interests but that doesn't change the fact that every time we meet Inuit from Canada or Alaska, well, it's like meeting your cousin. It is comfortable, you can recognize each other, you can feel the, I don't know how you call it, but you can identify immediately with each other.

This ability to maintain a bond between the widely dispersed Inuit has been critical to the ICC's aim to speak with one coherent and unified voice. But what does this voice say with regard to Arctic governance? What are the assumptions and challenges underlying the Arctic imaginary of transcendent nationhood?

The ICC: Creating a Viable Platform

In their local struggles for more rights across the Arctic, Inuit leaders' immediate action has been to question the southern population's right to impose its will on them. With the creation of the ICC these local Inuit struggles have been formulated into a platform that can be effectively presented and pushed on the international stage. The push to establish an explicitly international body was grounded in a number of realities. Among the Inuit, who are a trans-state people, there was a recognition that state borders work to divide and diminish the power of their small population. Related to this reality, legal decisions being made at this time regarding the rights of indigenous peoples were strictly state-centric, thereby limiting the

Inuit's struggles to the domestic realm, even though many of their problems were shared across borders. The Inuit leadership thus consciously positioned itself early on as advocating a comprehensive approach to Arctic governance that would diminish the power of political boundaries. This demand not only focused on how states deal with the Inuit but generally on how policies of development and environmental protection would be executed in the Arctic. The Inuit have thus long supported the internationalization of the region.

At the heart of the ICC's platform has been the concept of self-determination, a complex and somewhat indefinite term that entails the demand for social justice, cultural integrity, sustainable development and self-government. The ICC has become increasingly involved in international work geared towards ensuring indigenous rights, which has provided the legal foundation for demanding self-determination, while at the same time setting up country offices to help achieve these demands. A key argument in its demand for self-determination has centered on the Inuit's relationship to the natural environment. Sustainable development soon became the lynchpin of the ICC's insistence that the Inuit must be an integral part of any state or international Arctic policy.

The reason for this heavy focus on the environment was the advent of oil production in Prudhoe Bay, a development that deeply troubled the Alaska Natives of the region, but one that also offered significant opportunities. Eben Hopson, mayor of Barrow and the first president of the ICC, realized the need to position the Alaska Natives in the debate and the decision-making process surrounding oil development. He chose to focus on the Alaska Natives' historical connection to the land and hence to position them as the natural stewards of the region, which he could connect to the continuing land rights struggles occurring in Alaska, Canada and Greenland.

Political scientist Jessica Shadian, who has extensively studied the emergence and tactics of the ICC, has aptly argued that by focusing on the concept of 'stewardship' Hopson was carefully choosing a term that would tap into a particular stereotype about indigenous peoples, and Inuit in particular, as living in tune with nature by pursuing a subsistence lifestyle. The fact of the matter was, however, that the Inuit's reality was quickly changing, and, although many still engaged in hunting and fishing activities to supplement their household intake, an increasing majority lived in permanent settlements and either engaged in wage work or received assistance from the government. In these settlements, poverty was a significant problem and hence Inuit leaders were confronted with the challenge of how to achieve positive development.

To re-establish the old ways was not a viable option: many Inuit had lost touch with the survival skills needed for a nomadic subsistence livelihood in the Arctic. The increasingly urban Inuit, furthermore, many of whom had been acculturated to Western values, had aspirations that did not fit into a vision of returning to the past. In an interview, a Danish government official crassly remarked about his experience of living in Greenland:

> No one believes that you can be a hunter. I mean no one. It's impossible. You will die. I mean the coastal Inuits have a thousand years of history saying that you, from time to time, you will starve to death. But I mean no one in their full senses can want to live a traditional life like that. And no one in Greenland wants it.

Perhaps more importantly, however, most Inuit leaders realized that opting out of the development game in the Arctic would simply mean that southern populations would develop the region without the Inuit's interests in mind. In many cases, this would mean the destruction of natural resources on which the Inuit would have to depend. Hence Hopson was forced to recognize the potential of the oil finds on Inuit lands. In his words,

> I am not against oil and gas development in the Arctic. We Inupiat have gained great financial and political strength because of [it] .... [Although] we in the Arctic are not happy about [it] ... we understand that this development is necessary, and I, for one, want to cooperate closely to ensure that this development is done right. We can all benefit from development as we deliver to America her wealth that lies in our land, and beneath our seas.

This opening up to externally driven development is in significant contrast to the broad stereotypes that persist to this day of the Inuit subsistence lifestyle. Such a view, which pertains to indigenous people the world over, has, for example, been propagated by some in the environmental movement. Yet, for the Inuit, such a stereotype is double edged. For one, the image of Inuit living an existence of subsistence hunting and fishing may be one of sustainable living, yet the dark underside of this image is that of a backward and unsophisticated people, which is precisely the image that was called upon when justifying the subjugation of the Inuit and their lands to distant colonial, and later national, interests. Furthermore, the Inuit, who for better or worse do now possess and utilize the technologies of the twenty-first century, and who have become dependent on a money-based, market economy, are engaged in natural resource harvesting that goes beyond purely subsistence uses.

This disjuncture between the image and reality of Inuit ties to the environment can be seen particularly clearly in the ongoing debate over seal hunting. The Inuit have hunted seals longer than any other people in the world. Although there are few Inuit who live entirely outside the market economy, hunted seal meat still serves as an important supplemental food source and seal skins are used to make clothing and boots regularly worn by the Inuit. In addition to these practical uses, the seal hunt and the use of seal products is deeply ingrained in Inuit culture. Yet beyond these traditional uses, the Inuit view seals as one of the few living natural resources that can earn them some income, either through selling the pelts or the crafted products made of these pelts. There is also the possibility of an emerging market in the use of seal fat for extracting healthy omega-3 fatty acids. Yet this side of the Inuit's use of seals clashes with the image of them as self-sustaining and far removed from industrial society.

Such a clash has come to a head with the EU's ban on the importation of seal products. The European Union, following similar measures imposed by the United States earlier, has outlawed the importation of any seal products, much to the ire of Inuit communities around the world. In its defense, the European Union points to the inclusion of an exemption clause that is intended to allow Inuit to continue seal hunting for subsistence purposes. The EU regulation reads as follows:

> The placing on the market of seal products which result from hunts traditionally conducted by Inuit and other indigenous communities and which contribute to their subsistence should be allowed where such hunts are part of the cultural heritage of the community and where the seal products are at least partly used, consumed or processed within the communities according to their traditions.

In other words, the European Union is here still strongly committed to its vision of the Inuit as purely traditional hunters engaged in culturally significant subsistence activities.

Such a view, however, is not how the Inuit truly see themselves, and it flies in the face of the Inuit's desire to hunt seals to make a profit in a region where work opportunities are few and far between. As a leader of ICC Greenland emphatically stated:

> Never will Inuit only continue to use kayaks as many environmentalists want them to do. So if you're asking will they pledge to stop all development, other than development that happened a hundred years ago, like the EU seal ban wants them to do? No.

And yet, the idea of the Inuit as natural stewards of the land is a notion that the Inuit have continued to promote with great success. The ICC's message of Arctic stewardship started to resonate fully with the end of the Cold War, when the priority of security dissipated. With the new geopolitical reality of the post-Cold War world, different issues were emerging that revolved around the economic opportunities and environmental constraints in the Arctic as well as the Arctic's increasingly apparent role as a bellweather of climate change. These were issues toward which the Inuit were well placed to take part.

On the heels of these reflections, however, also came nagging questions about sovereignty. A number of critical issues still remained on this front: border disputes, questions regarding state claims based on the extension of the continental shelf and whether the Northwest Passage should be deemed internal waters or an international strait. These critical questions became even more pronounced when the thawing of relations between the East and the West was accompanied by a thawing of the ice in the Arctic region, largely attributed to global warming. In the eyes of many, such warming could only mean even greater opportunities in the North in terms of accessing the vast amount of mineral and petroleum resources believed to be located there as well as the opening up of shipping routes across spaces that once were inaccessible due to the ice.

Jessica Shadian has made the convincing case that in this environment it was Canada, in particular, that saw an opening by engaging and cooperating with the Inuit. ICC Canada had developed a particularly strong voice via the engagement of Inuit leader Mary Simon, which meant there was increasing contact and collaboration between the ICC and the Canadian federal government. For its part, the ICC had been working on a comprehensive Arctic policy for more than a decade, the hallmarks of which were a focus on internationalism and environmental sustainability. Under the influence of the ICC, the Canadian government hence initiated a process that would eventually lead to the establishment of the Arctic Council in 1996, which today has become the primary international forum for setting international policy agendas for the Arctic.

Once established, the mission of the Arctic Council was, in its own words, 'To enhance Arctic environmental protection while promoting sustainable economic development, to further empower Arctic aboriginal peoples at the domestic and international levels, and promote regional security'. By helping to establish such a high-level international consultative body with this mission, the Canadians saw an opportunity not only to present themselves as committed to the environment, particularly in the North, but also as being on good terms with its indigenous peoples. In fact, the Inuit were becoming an integral aspect of the 'northerness' that continues to serve as a crucial focus of Canadian identity.

For the Inuit, the creation of the Arctic Council, which the ICC strongly supported, was a major breakthrough. The internationalism which they had practiced within the ICC, and for which they had been advocating in an attempt not to be overrun by domestic agendas, had come to fruition. Not only that, but the ICC, which was given the status of permanent participant in the Arctic Council, was now empowered to sit at the table, almost as equals, with states, along with five other permanent participants representing the other indigenous peoples of the North.[2] Lastly, the focus on environmental sustainability in the Arctic gave the Inuit – who had been successful in projecting a self-image of being natural stewards of the land – a secure reference point for demanding further participation.

Today the Inuit are seen as key stakeholders in the Arctic. As one US State Department official involved with Arctic matters frankly stated, 'The ICC is definitely a force to be dealt with in the Arctic. There is no question about it.' Yet the Inuit see themselves as more than just stakeholders. They may have managed to get their seat at the table through a very opportune framing of themselves, but deeper than this is an unshakable sense among the Inuit that the Arctic represents their domain, that they know it best, and that they are most affected by anything that occurs there. Hence, in the view of the Inuit, it is only natural that they be made an integral part of how the Arctic develops. Having been made a permanent member of the Arctic Council, the ICC now has a stage and an important audience where they can bring attention to their needs and concerns. And Inuit leaders have many requests and demands: to play a greater role in how resource-extraction activities will be conducted, to safeguard natural hunting grounds, to maintain freedom to harvest and sell natural resources, to influence the direction of research in the Arctic, and to have indigenous knowledge be incorporated into the 'scientific' endeavors that will shape future policy.

Having gained a strong representation in the Arctic Council, the ICC has continued to position itself as a natural partner in the governance of the Arctic. As one ICC Canada official declared,

Even if the position of a certain state is that the Inuit aren't legally mandated to be at a table on a certain issue, it makes sense to have the Inuit at the table; sometimes you have to be practical .... Even the discussion on the construction of ships, we should be there for that. Twenty years ago, that would have been laughable, but I think the Inuit have a lot to offer about things you may not even know about.

Indeed, the Inuit have a well-deserved self-confidence, rooted in their long tenure in the Arctic. They have furthermore used this position to make themselves 'a force to be reckoned with' when it comes to the future of the Arctic. In particular, they have been successful at promoting an imaginary under which the Arctic, while under the political control of sovereign states, is also characterized by identities and values that transcend state borders. Yet, despite these tremendous achievements, the Inuit can by no means rest on their laurels. Outsider interest in the Arctic has grown dramatically the world over and, in the end, the Inuit are still a very small population facing many internal, as well as external, challenges. Some fear that now, when it appears that the Inuit have achieved a secure and powerful voice in the Arctic, that voice is in danger of being silenced again.

Internal and External Challenges

As was discussed in Chapter 1, the Ilulissat Declaration of 2008 can be seen as an affirmation by the five Arctic Ocean coastal states – Canada, Denmark, Norway, Russia and the United States (the Arctic Five) – that the future of the region will reproduce the centrality of the sovereign state as the fundamental political actor. The declaration asserts that the maritime portions of the region will be governed according to the rules established by UNCLOS. Under these rules, although individual sovereign states may claim some level of jurisdiction over resources in portions of the Arctic Ocean, the fundamental legal status of the ocean, beyond the narrow 12 nautical mile coastal strip, is beyond state sovereignty. Yet, when the signatories asserted, 'We [the five states] will keep abreast of the developments in the Arctic Ocean and continue to implement appropriate measures' they were implying that even distant portions of the Arctic Ocean beyond 350 nautical miles from shore were somehow within their sphere of influence. The five Arctic states, individually and in cooperation with each other, have attempted to demonstrate this influence through military investments in the region and through active, nationally sponsored polar research programs.

The state-centric focus of the Ilulissat Declaration – both the convening of the conference outside the Arctic Council framework, which meant that permanent participants were not invited, and the text that insinuated that it would be states and no one else who would keep abreast of developments and formulate appropriate policy for the Arctic – was immediately viewed with alarm by the ICC. As one ICC official explained with regard to the Ilulissat meeting:

As you know they did not invite indigenous people, they did not invite Inuit specifically, who are also very much bordering the Arctic Ocean, to speak or to be present at negotiations in any kind of way ... like in the Arctic Council .... So first of all, Inuit were excluded. Secondly what the declaration said was quite unacceptable to Inuit. It talked about the importance of international law and international instruments but it didn't talk about all the international instruments referring to indigenous rights, human rights of indigenous people.

The ICC had already begun to deliberate on issues of sovereignty prior to the Ilulissat summit, but it was the Arctic Five meeting that prompted it to issue its own Declaration on Sovereignty in the Arctic. In this document the ICC does not engage in a fully fledged dismissal of the concept and institution of state sovereignty but it does question the simplicity and matter-of-factness that is often assumed when issues of sovereignty emerge. In this document, the ICC proceeds to present the Inuit multifariously as a people with universal human rights, as citizens with defined state-sponsored rights, and as an indigenous people with specific rights within these states. The ICC's main objective in this document is to affirm the role of the Inuit people as an integral component of any possible sovereignty arrangement in the Arctic.

In its entirety, the ICC's Declaration on Sovereignty in the Arctic actually takes a dual approach – on the one hand questioning the contemporary authority of the concept of sovereignty and, on the other, pointing to the integral role that the Inuit play in its execution. With regard to the first point, the declaration mentions the 'contested nature' of the notion of sovereignty and highlights the lack of any fixed meaning that the term may hold. Next, the document works to position the Inuit as integrally fused with the Arctic's land, sea and ice, and thus it indirectly questions the privileged status of distant capitals to exert their power over the Arctic region. By highlighting the way in which Inuit livelihoods cross borders (between states but also between land and water), the ICC is challenging the fundamental basis of territorial sovereignty in the control and ordering of space. This assertion supports the ICC's broader argument that the application of state sovereignty to the North requires modification to account for the ways in which the region is actually used, a point that we considered in greater detail in Chapter 3. By rejecting state sovereignty as an absolute concept – either you have it or you don't – the ICC manages to avoid playing the political game fully on the states' terms and instead succeeds in highlighting the Inuit demand, and indeed their perceived right, for self-determination.

In some ways, then, the concept of self-determination is even more radical than a call for an independent sovereign state, as it demands the right to opt out of the power structures imposed by any external sovereignty arrangement. Critically, the tactics involved in such a political stance, at least for the ICC, entail a radical focus on identity – in this case an Inuit identity that is placed squarely outside the state mechanism of assimilation. The desire, and perhaps need, to present an identity squarely outside western (or southern) norms, is something that was clearly expressed at the ICC's 2010 General Assembly in Nuuk, in which the distinctive culture of the Inuit was driven home and celebrated in all its forms, ranging from matters of spirituality to indigenous knowledge. This self-presentation, as existing within, yet at the same time beyond, nation-state boundaries, underpins an underlying vision within the ICC of a people who transcend the division of societies into discreet bounded spatial units.

Yet, it cannot be forgotten that the ICC continues to be a political actor with limited power, and if push came to shove it would have much to lose if states were to decide simply to circumvent it. For instance, the political principles that the ICC favors, rooted in a number of UN declarations and other conventions, have to date been largely questioned if not flatly rejected by a number of powerful states. The United States has yet to ratify the UN Declaration on the Rights of Indigenous Peoples, and Canada has only just recently signed it. It is with regard to this reality that the ICC has also woven into its Declaration on Sovereignty in the Arctic a further layer of meaning that seeks to sediment the importance of the Inuit people within a more traditional, state-centered system of territorial sovereignty. The brunt of this approach is pointing out that it is the Inuit and other indigenous peoples of the Arctic who give states the basis of claiming Arctic lands due to the historic habitation of their citizens.

When questioned about this juxtaposition between a more and less forceful message within the ICC's declaration, the ICC official quoted earlier stated quite frankly, 'Even though they [the Inuit] see these boundaries that you are looking at as artificial, ultimately they do have a certain respect for them. It may be in an unfortunate way, but we can't just erase history'. It is important to note that the Inuit, particularly in Canada, have a significant history of being used as pawns in state sovereignty claims that have hinged on sufficient occupation of land, most obviously with the Grise Fiord relocation, but in other instances as well. The ICC has thus carefully crafted its unique position into a demand for greater representation within any state-centric approaches to Arctic decision making. As the ICC Declaration on Sovereignty in the Arctic states, 'The foundation, projection and enjoyment of Arctic sovereignty and sovereign rights all require healthy and sustainable communities in the Arctic. In this sense, "sovereignty begins at home".' In other words, the ICC is making clear that any state-based sovereignty claims that may be made ultimately depend on the presence of functioning habitation in these regions, and this habitation is provided, first and foremost, by the Inuit. Furthermore, to make this habitation successful, and to enable it as a foundational pillar of sovereignty claims, it needs to be supported through effective governance (and, presumably, investment in social and physical infrastructure).

Hence, with regard to pre-existing and stubborn conceptions of territorial, state-based sovereignty, the ICC's central point is that state-based sovereign control of the far flung territories of the Arctic can occur only with the incorporation and respect of the Inuit's customs and their region-specific knowledge. In this way, even if the ICC's more radical departures from the traditional state-centered system are bypassed by the Arctic rim states in their scramble to lay claim to the Arctic Ocean's resources, the ICC will still succeed in placing the Inuit people as pivotal and integral players in this process.

Again, with respect to the radicalness of the declaration, one ICC Canada official interviewed concluded, 'If you look at the last paragraph, it's about holding hands with government. I mean how unradical can that be, and so it's practical.' Thus, despite its questioning of the boundaries that dissect the Arctic into separate sovereign realms, each at the mercy of a distant southern capital, the ICC sees the need to 'hold hands' with government. But even here it seeks to do so in a transnational way. From its very inception, the ICC has pushed for a regional approach to the Arctic that transcends the limited interests of single nation-states. The Arctic Council is, in many ways, a model of this vision as it not only brings the various states to the same table in a spirit of cooperation but it also allows for the significant input of non-governmental organizations, such as the ICC. Even though the Arctic Council is a policy-shaping institution with very limited policy-making authority, it is viewed by the ICC as making an important contribution to shared governance in the Arctic.

The Ilulissat Declaration, on the other hand, was seen by the ICC as representing a step back to narrow national (or, at best, multilateral) interests, thereby doing a run around the Arctic Council, and the influence that indigenous peoples have in it via the ICC and the other permanent participants. Since the Ilulissat Declaration was signed, the Arctic Ocean coastal states have all reiterated their support for the Arctic Council and, indeed, the Arctic Council has taken on an increasingly important role in not only discussing policy but in actually setting up legislation. But this is not necessarily good news for the Inuit. In fact, the ICC is openly concerned that the Arctic Council is also in danger of cutting out the influential role that the Inuit have there.

As a governing body that actively seeks out the input of the Inuit via the ICC, the Arctic Council was a very positive development for the Inuit, one that they themselves worked to make happen. The crux of this influence lies in the fact that the ICC is designated with 'permanent participant' status, a position reserved for only a handful of Arctic indigenous groups, of which the ICC is, by all accounts, the most powerful. Being a permanent participant does not give the ICC voting rights (the Arctic Council functions on the basis of consensus reached by all eight member states), but it does allow the ICC to take part in most of the agenda making and in the deliberations that lead up to decision making.

Other non-members that may take part in Arctic Council meetings include the permanent observer states, of which there are currently 12, and ad hoc observer states, which need to apply to participate in each meeting. With all of the interest in the Arctic as a region, and with the growing importance of the Arctic Council, a number of states, and even NGOs, have been scrambling to obtain permanent observer status (see Chapter 8). It is this potential of changing dynamics within the Arctic Council that concerns the ICC. As one ICC official explained,

> The Arctic Council is a groundbreaking institution, and there is a little bit of a concern that as new observers come in from all over the world, that's okay, as long as the role of ICC is not watered down. But there is a status there that is unique. . . . There is a concern that the Arctic Council may change too radically.

There is also concern regarding the potential strengthening of the Arctic Council to achieve more policy-making capacity. When asked about these developments, another prominent ICC leader stated: 'That's what we hear, that they want to strengthen some areas. But I'm afraid when they're saying it that they want to keep out the indigenous peoples.' There is thus a very clear sense that the rapid developments and burgeoning interests in the Arctic are viewed by many Inuit groups as endangering the gains they have made to this point. The aim of creating a transnational and less state-centric space in the Arctic is being challenged by growing national interests that are finding expression both within and outside the Arctic Council. Yet what is the solution? Although the concern of the Inuit is understandable, several sympathetic outside observers are skeptical of the ICC's current stand.

One activist in the Inuit rights community explained:

> The long-term interests of Inuit, of all Arctic indigenous peoples, is one of engagement, because people are coming in whether you like it or not. The world is changing whether you like it or not, so you want to strengthen the Arctic Council so that you can exert influence more

broadly, not just nationally or in the circumpolar world, to use the Arctic Council to influence the broader agenda. And that, unfortunately, is not the way that many of the ICC politicians in the last little while have seen matters.

He then continued to emphasize the importance of the Arctic Council to the ICC:

Basically it is a forum for them to amplify their voice. You couldn't get an audience with the government of China if you went to them yourself, but suddenly you're sitting there with the US, Canada and Russia respectfully asking your opinion on an issue ... I think that what has happened right now, to be honest, among the permanent participants, and this is no denigration of the people who are involved, they are all friends and colleagues of mine, but they have atrophied.

Atrophied. This is certainly not a good state to be in when everything around you is abuzz with change. So much progress has been made by the very people who still now are the key voices in the movement. Are they losing their edge? Is the leadership ageing and out of ideas? At the General Assembly in Nuuk, it was noticeable that the leaders all seemed to be elders. The new ICC president following the Nuuk conference, Aqquluk Lynge, was an old hand who had been with the organization for decades. The nostalgic speeches of the good old days served as the glue between the various guest speakers, who spoke of contemporary problems. It was as if the identity of the sustainable hunter and fisher that was being rehashed was kindled mostly via the memories of the older Inuit generation and the exotic imaginaries of the non-Inuit who were present.

Listening to these speakers, and embracing the mood of the event, however, it was difficult not to wonder about the continued effectiveness of the ICC's approach. Was this simply nostalgia or denial? One speaker, for instance, questioned the scientifically grounded warnings against eating seal meat that, in some areas, has shown an increase in contamination levels. 'Are we Inuit expected to stop our way of living based on a science that is not ours?', he seemed to be asking. 'How do we know that our ill-health doesn't come from the western foods that we increasingly eat?' 'Perhaps we should stop eating so many canned beans!' he then suggested to peals of laughter. Indeed, western foods may not be healthy for the Inuit but is it wise to deny the findings of western science, even if those findings problematize a traditional life that is increasingly imaginary?

6.2   Traditional seal use demonstration associated with the General Assembly of the Inuit Circumpolar Council, Nuuk, Greenland.

The tension here between the identity of the archetypical Inuit of long ago and the modern Inuit, living within the context of an increasingly urban and industrialized life, was palpable. This tension, of course, has been there since the very emergence of the ICC, when still relatively traditional cultures were faced with the prospect of a regional oil boon. Eben Hopson knew that this would be a difficult identity to balance, resisting full assimilation and the loss of one's culture while at the same time adapting to and benefiting from what the encroaching southern culture has to offer. Today the oil question is still a major one, as many Inuit lands could still be opened up to oil exploration and extraction. Should oil development be embraced as a cash cow that can be used to reinvent the Inuit as the modernized, urban gatekeeper to the North? Or should it be resisted, and, if so, on what basis and with what alternative in mind?

There are different views on this issue within the Inuit leadership. Aqquluk Lynge, a Greenlander and the current ICC president, has remained skeptical, much in contrast to the Greenlandic government's position that was elaborated in Chapter 4. Certainly, an oil find in Greenland would dramatically change Greenland's reality, bringing in a swarm of outsiders seeking to benefit from the newly established oil economy. The concern here is that the Inuit's quintessential identity would be lost forever if the oil economy were allowed in. The alternative would be to improve what is already done: a focus on mixed economies where renewable resources are the main source of income as well as supplying a culturally significant supplement to food needs. A greater focus on education and the subsequent hiring of indigenous peoples by the dominant government sector is also a part of this equation. The reality remains, however, that without significant revenues coming in from non-renewable resources (i.e., oil and gas), there will remain significant dependence on outside support.

As a consequence of these challenges, there have emerged noticeable rifts regarding the resource issue among the Inuit leadership. As one leader in ICC Greenland explained with regard to Greenland's oil ambitions: 'The ICC position has always been that we would not allow offshore drilling .... They [Greenland] have decided that they will go ahead and, of course, it is with great regret that we see that.' Yet, the ICC position in the United States and Canada, as witnessed by statements made by ICC founder Eben Hopson, was never categorically against oil drilling. Thus, in light of the oil extraction that has already occurred in Alaska and Canada, Greenland remains the last 'unspoiled' land. Nevertheless, when the question is raised whether to develop oil or remain dependent on Denmark there is a resounding answer in Greenlandic politics, with a fully committed preference for the former, or rather a desire to overcome the latter by any means.

The paradox for the ICC is that it has itself benefitted from oil. The North Slope Borough, home to Eben Hopson, is fully funded by oil, and there is no doubt that the Borough has been a major financial sponsor of the ICC. This is certainly not something that the ICC wants to have get in the way of its unified image. When asked about the potential rift between ICC Greenland and its anti-oil stand, one Canadian Inuit leader dismissed any suggestions of disagreement:

> It surprised me that the media said, 'Oh there are varying degrees of agreement.' No, I'm sure that was still the consensus: No exploration ... until the proper safeguards are in. ICC Greenland has been saying the same thing all throughout .... Nothing has changed. The only thing that has changed is the interpretation that the media [presents].

The consensus to which the interviewee was referring was made official in a document entitled the 'Circumpolar Inuit Declaration on Resource Development Principles in Inuit Nunaat'. This document declares that in addition to traditional renewable resources, 'Responsible non-renewable resource development can also make an important and durable contribution to the well-being of current and future generations of Inuit.' The document then spells out how this development must be conducted in a sustainable fashion so as not to disrupt the more traditional activities of the Inuit. The document also notes that Inuit partners should be included in any development plans. With regard to revenues, these must first be directed to benefit the inhabitants of that region, i.e. the Inuit. Thus, as in the past, the Inuit leadership is attempting a balance between tradition and modernization, between subsistence and industrialization, between environmental concern and economic growth. Up until now, the ICC has been very successful with this formula. Yet as the pressures mount to increase access and exploitation of the region, the Inuit will undoubtedly continue to feel the winds of change. Some may benefit, others may not.

In Nuuk at the ICC meeting, a film festival screened a film called *Inuk*, about a troubled Inuit teenager from Nuuk who is sent to the North for rehabilitation by getting in touch with the old traditional Inuit ways. When he successfully hunted his first seal the mostly Greenlandic Inuit audience cheered in ecstasy. Here was an urban youth gone astray tapping into his 'authentic' Inuit identity. Yet just a few months earlier, the very same Nuuk inhabitants could be found waving Greenlandic flags on the street, welcoming a delegation of Alcoa executives who had come to lobby for the opening of an aluminum smelter in Maniitsoq. Industry in Greenland would lead to revenues, which would mean greater independence from Denmark. Yet the Inuit wage workers in the smelter, if, as has been suggested, imported Chinese do not get the jobs first, would be placed even further out of touch with that mythical Inuit identity of the North. It seems, at some point, that something has to give. When it does, the ICC will probably have to take a side, either choosing greater assimilation, with its focus on citizenship and the acceptance of a state-centered Arctic, or continued resistance on the basis of an imaginary that inherently questions the validity of the nation-state order and the economies that come with it.

7.1   Polar Bear, Bering Sea, Alaska.

CHAPTER 7

NATURE RESERVE

The threatened polar bear is probably the most haunting and enduring image of the impact of climate change in the Arctic. The species is dependent on sea-ice, which it uses for sleeping, breeding and hunting. Polar bears will wait on sea-ice for seals – their favorite prey – and then hunt them when they surface. They need this ice because, despite being powerful swimmers, they are not themselves aquatic and the ice provides the necessary platform for their hunting.

Unfortunately for the polar bear, sea-ice has been in relatively shorter supply lately. With climate change leading to less sea-ice, polar bears are left to forage from land, which is not as effective for seal hunting. Diminished seal consumption leads to reduced body mass when it is needed in the winter months, subsequently threatening their lives and leading to underfed cubs. Increased habitation on land instead of ice also leads to greater contact with people, which can lead to bears becoming desensitized to humans and thus endangered in a different but no less lethal way.

In short, the concern is that reduced pack ice is leading to fewer and less healthy polar bears. Although it can be difficult to determine the overall health of polar bear populations, changing Arctic conditions do not appear favorable for these northern animals. Concerns about the health of the species led the United States in 2008 to designate the polar bear a threatened species under the 1973 Endangered Species Act. The United States has also pushed for the polar bear to be listed under the major international convention dealing with endangered species: the Convention on International Trade in Endangered Species of Wild Flora and Fauna (CITES), a proposal that was rejected at the CITES 2010 Meeting of the Parties.

Nonetheless, the image of a lone polar bear clinging to an ever-shrinking ice floe remains the ubiquitous depiction of an endangered North. The threatened polar bear is seen as something like the 'canary in the coal mine' – a creature that will be one of the first to show signs of suffering from environmental change. But it is also a popular photogenic animal – beautiful and majestic. As such, the polar bear in some ways has come to represent the idealization of the entire Arctic – free, powerful and wild.

The case of the polar bear in many ways typifies the final imaginary that we profile in this book: that of the Arctic as nature reserve (or national, or international, park). Under the most extreme version of this imaginary, the Arctic is conceptualized as having largely escaped the sullying hands of humankind. In this particular vision, the Arctic is a place in need of protection from those who would seek to utilize its resources, as detailed in Chapter 5. Here the Arctic is less a place for humans, who are – to varying degrees – integrated into global political and economic relations, and more an isolated region for the (non-human) creatures that live there.

This vision, of course, is extreme, and few people, if asked, would say that the Arctic should be preserved just for wildlife to the exclusion of humans. But the caricature of a pristine and isolated Arctic is present in Arctic debates in much the same way as the other imaginaries presented here: it is rarely advocated in its purest version but it nonetheless has a tendency to influence debate. The 'nature reserve' imaginary can be especially powerful when the idea of a pristine and wild Arctic is posited against one of development and economic growth.

But does this binary – beautiful nature and wild animals, versus a resource frontier on the brink of exploitation – really capture the circumstances of the Arctic? And do the people who would like to protect the Arctic's wild beauty actually expect to exclude the people who live and work there?

The Vision

For many who have never lived in or even visited the Arctic, the image that may come to mind when thinking about the region is a cold white world of polar bears, narwhals, seals and icy waters. This Arctic is a remote place and the four million people who actually live in the region are largely absent. Or, if they are not absent, they are reduced to anthropological caricatures, mystically connected with – and a part of – the nature that surrounds them. This is a necessary caricature as the imaginary of the Arctic as pristine rests on the ideal that there are no nature-disturbing people there.

Currently, much of the concern about the future of the Arctic revolves around climate change. In 2007, satellite photos of the circumpolar Arctic captured images of unprecedented summer ice melt, leading many climatologists and other scientists to worry about the acceleration of temperature change and sea-level rise. Their concerns were confirmed by the record-shattering 2012 ice melt, which was worse than the one experienced just five years earlier. As was discussed in Chapter 1, melting ice does more than affect polar bears. Ice and snow have a high albedo; that is, they reflect a high proportion of incoming solar radiation. They act like mirrors in a sense – sunlight is reflected back towards the atmosphere. The darker waters of the Arctic Ocean, however, have a much lower reflectivity, leading to increased absorption of the sun's warmth. Eventually, these waters become permanently warmer, creating a feedback loop in which the warmer water consequently melts more ice, producing more water to absorb the sun's energy and thus raising the temperature of the water further.

Because of the ecological sensitivity of the Arctic, the effects of climate change can be seen dramatically there. But changes in the Arctic itself are also important drivers of climate change. Although melting sea-ice contributes relatively little to sea-level rise directly, land-based ice sheets do make a significant contribution as they melt. Also, the temperature and salinity impact from melting ice can greatly change ocean current patterns and affect marine creatures as their migration routes adapt to changed conditions, eventually leading to displacement of the species that has resided in a given area. Furthermore, ocean currents have a significant impact on adjacent land-based climate systems (e.g., the role of the Gulf Stream in moderating Europe's climate, despite its relatively high latitude), so these changes in ocean currents can affect terrestrial ecosystems as well. And finally, turning back to the Arctic, sea ice protects communities from coastal erosion, and the warmer temperatures overall can result in seasonal melting of the permafrost on which the structures in far northern communities are typically built.

Nonetheless, most residents of the North distance themselves from the 'nature reserve' imaginary. Their resistance is not based on a lack of understanding of the threat that climate change poses for their livelihoods. Indeed, as a senior US Arctic research official noted in an interview, some of the strongest advocacy for climate change research has come from Republican elected officials from Alaska, notwithstanding the national Republican Party's skepticism of climate change science. Likewise, over the course of our interviews with Arctic residents, numerous respondents – both indigenous and non-indigenous – referred to the region's varied and unique nature and the need to protect it for residents' continued use and enjoyment. Rather, Arctic residents' resistance to the imaginary is rooted in their fear that the preservation of nature will come at the cost of badly needed economic development or the traditional use of living resources. They see a need to balance these two concerns – protection and development – and believe that focusing too much on one over the other could endanger their continued well-being.

To Protect, Use, or Both?

Even though the 'nature reserve' imaginary of the uninhabited, wild Arctic is rarely, if ever, advocated in its extreme form, there is considerable debate regarding what is the best course to take so as to ensure the greatest measure of wilderness preservation without limiting the ability of the people who live in the Arctic to benefit from its natural resources.

Environmental non-governmental organizations, or ENGOs, are probably the groups most associated with the promotion of environmental protection in the Arctic. Their work throughout the circumpolar North attempts to offer what they see as a balance between development and the environment. Thus, their goal has not been to oppose all human activities in the Arctic but rather to discourage those that cause the most environmental harm and to encourage development that can be done in an environmentally sensitive manner. A Canadian ENGO representative explained his group's strategy as follows:

> [T]here's a number of areas where we don't think there should be any development. However [we have] not said we are against development in that area at large.... [We want to know,] What does an alternative economic vision of the North look like, one that includes resource development but makes sure that the benefits of resource development accrue to local peoples, more so than it does now, and that has a place there for the economic benefits for conservation?

But it is not immediately obvious how best to achieve this goal. ENGOs and others want the Arctic to have protection – but what kind of protection is most necessary and from what activities? In this regard, ENGOs and others have both theoretical and practical differences of opinion. These differences can be grouped in large part by their relationship to the idea of the Arctic as a threatened wilderness.

Some ENGOs propose that nature be protected against most development, and many of these would advocate setting aside large portions of the Arctic as a 'national park' or 'World Heritage Site', or even establishing an international treaty similar to the Antarctic Treaty System that governs the Antarctic region. A solution along these lines would provide the greatest amount of local and regional environmental protection, but it is the least popular solution among the residents of the North, the companies that wish to develop Arctic resources and the policy makers whose support would be absolutely necessary to bring such an objective about.

Other ENGOs prefer to work with indigenous groups to protect the region from potentially destructive oil and gas development, but these partnerships can be uneasy and often difficult to manage. ENGOs and indigenous groups share some similar goals: the protection of wildlife and habitats are issues on which the two typically agree, as they also do regarding efforts to monitor the impacts, and reduce the production, of globally generated pollutants that have particularly negative impacts on the health and livelihoods of Arctic people (e.g., greenhouse gases, persistent organic pollutants, short-term climate forcers). However, the importance of traditional foods in indigenous cultures and the subsistence lifestyles necessary to obtain them are issues that can cause disagreement. Some environmental groups deplore the hunting of seals and whales, in which indigenous peoples of the Arctic have participated for centuries, and this has sometimes stymied efforts to spur cooperation between these two interest groups. Moreover, although the situation certainly varies from case to case, in many instances there are significant differences in the relative power of the indigenous groups and ENGOs who attempt to engage in partnerships – for example, in access to funding and media attention.

Regardless of whether ENGOs take a narrow preservation approach or consider sustainable and precautionary development appropriate, they generally have disagreements with those who wish to extract resources from the Arctic. These may be limited to disagreements over specific behaviors undertaken by extractive industries – petroleum, natural gas, or mining – or they may also involve differences of opinion with the national, regional or local governmental authorities that are tasked with regulating uses of the Arctic lands and waters that the ENGOs seek to protect. Additionally, the extent and locus of government involvement in environmental protection and/or Arctic resource development differs significantly across the Arctic states, as does the extent to which extractable resources are present throughout the wider region.

Petroleum tends to be the resource associated with the Arctic that is most discussed in the media, and drilling for oil remains a dominant concern both for those interested in extracting resources and for those concerned with that activity's environmental hazards. However, other opportunities, such as mining and natural gas development, are also prevalent in some areas of the Arctic, and these industries pose their own considerations. Thus, ENGOs face diverse opportunities and challenges across the Arctic, as they pursue goals and strategies that must be tailored to the particular region where they are to be implemented.

'Purity' of Vision: A National Park or International Treaty

'The real focus from a good section of the country has been to make all of Alaska into one park, and that ongoing effort, challenge if you will, continues,' a representative of the state's mining interests opined in an interview. In his view, the environmental movement in the United States is seeking to rope off the Arctic and prevent his industry from any sort of meaningful development – in other words, to turn the state into an 'Alaska National Park'. He offered an anecdote from his experience of giving a talk at a forum where he was scheduled to appear alongside a representative of an ENGO:

> [There's this] fellow ahead of me from [an environmental organization] ... I'm listening to him and all of a sudden I learned that ... we're talking about an international park, we're talking about a World Heritage Site and a marine biosphere reserve.

For this mining industry representative, ENGOs are not small, relatively under-resourced organizations that are seeking to influence the trajectory of the state's economy. Rather, they are emissaries of a powerful international movement to declare nature 'off limits'.

Anxiety that the Arctic could become an 'international park' is not restricted to representatives of large industries in the United States: interviewees around the circumpolar region ascribed this desire to environmentalists and ENGOs. One Canadian consultant, who works with the federal government and indigenous groups on Arctic-related issues, claimed that, '[The Arctic seems to be] a wilderness for a lot of NGOs outside the Arctic: Put a fence around it, [create a] giant theme park, everybody is safe.' This consultant went on to decry the tendency of these groups to be 'uninformed' about the connections between the Arctic and the rest of the world and unconcerned about what he perceived to be the real issues and challenges in the region.

In fact, it became apparent throughout the interviews that few in the region truly feel that enclosing the Arctic to keep it 'safe' from human development is an achievable, or even desirable, goal. Indigenous peoples resent it, governments find it a troubling prospect, and environmental groups themselves feel like this is the wrong way to foster environmental protection in this complex region. Therefore, rather than pursue some sort of national park project within individual countries, some ENGOs promote an international treaty that will manage human activities in the Arctic. The model cited here typically is the Antarctic Treaty System, which entered into force in 1961 and governs activities in the southern polar region.

The Antarctic Treaty System was negotiated in the context of the Cold War. Seven states claimed part of the Antarctic continent (Argentina, Australia, Chile, France, New Zealand, Norway and the United Kingdom), and both the United States and the Soviet Union refused to recognize others' claims or relinquish the ability to make future claims of their own. Thus the Antarctic had the potential to become a Cold War battleground, a place of conflict between the world's two great superpowers should they decide to take an interest. The Antarctic Treaty put such fears to rest by mandating the Antarctic as a place of scientific cooperation, forbidding military presence or nuclear testing on the continent and calling for the protection of the living species of the region. The environmental provisions in the Antarctic Treaty were further strengthened decades later by a subsequent Protocol on Environmental Protection, which entered into force in 1998. This protocol called for the preservation of the environment as a fundamental consideration for any planned activities in the Antarctic region.

The Antarctic Treaty System has proven to be effective in preventing the Antarctic from becoming a zone of political or economic conflict. This is why it is often held up as a model for an 'Arctic Treaty' that would perform the same sorts of functions for the northern polar region. The Cold War is over, but concerns remain about the possibility of conflict in the Arctic. Though for the most part there is little reason to expect conflict in the near future between the Arctic states over the area, the discussion of such conflict and its potential remains a part of the broader debate on the Arctic's future. ENGOs, worried about environmental degradation due to the possibility of economic exploitation of the region and its natural resources, have at times pointed to an 'Arctic Treaty' as a potential solution that could respond to concerns over both conflict and environmental degradation.

Many of the major international ENGOs favor an Arctic treaty, at least as an ideal, even if they acknowledge that such a treaty might not be achievable. Following the model of the Antarctic Treaty System, the primary function of an Arctic treaty would be to promote peace in the Arctic, necessary for staving off a resource war. However, by its very nature such a treaty would also grant a degree of environmental protection to the area. Having the five Arctic Ocean coastal states, or the eight states with territory north of the Arctic Circle, come to an arrangement to defend an environment seen as particularly fragile would be a tremendous victory for the environmental movement. Such an agreement would by definition have to include the United States and Russia, moreover, making it an international convention backed by two of the most formidable powers in international politics. With two great powers behind this agreement, other, non-Arctic, states that have grown increasingly interested in the region (e.g., China) would be unlikely to challenge it.

One proponent of a binding legal instrument that would govern uses of the Arctic marine environment is the World Wide Fund for Nature (known as the World Wildlife Fund in the United States and Canada, or the WWF for short). In a detailed 2009 report, the WWF's Global Arctic Programme, which was then based in Oslo and has since moved to Ottawa, argues that because of gaps in existing regulatory regimes (e.g., those sanctioned by UNCLOS or enabled by the Arctic Council), a new comprehensive mechanism is required to protect the waters of the Arctic Ocean and the wildlife contained therein:

> The governance and regulatory regime that currently exists in the Arctic may have been adequate for a hostile environment that allows very little human activity for most of the year. But when the Arctic Ocean becomes increasingly similar to regional seas in other parts of the world for longer and longer parts of the year, adequacy cannot be assumed and reform of the regime is indispensable.

The WWF is far from alone in supporting a binding international governance instrument, acceded to by states, as the best possible means to govern the Arctic Ocean. Fellow ENGO Greenpeace, in a much shorter 2009 report of its own, calls on states to establish 'a permanent, equitable and overarching treaty or multi-lateral agreement'. Another ENGO, Oceana, does not explicitly call for a treaty in its 2008 report on the Arctic, but the report begins by pointing out that the world must act to save the region precisely because such a treaty does not currently exist to protect it.

To be clear, the proposals floated by these ENGOs and other environmental groups typically are not as extreme as they are often portrayed by their opponents: few proposals seek to cordon off the entire Arctic as an international park, marine sanctuary, or World Heritage Site. Perhaps in recognition of the political reality in which it operates, the WWF proposal (somewhat like the ICC Declaration on Sovereignty discussed in Chapter 6) takes pains not to impinge on the sovereign power of states over their land territory, restricting itself to the *marine* Arctic. Nonetheless, these proposals do reflect the general spirit of the 'nature reserve' imaginary by urging states to regulate the maritime Arctic as a place of special environmental protection.

There are some major problems with these proposals, however. Foremost is the concern that such a treaty would never be sanctioned politically. It would be impossible to gain state ratification and impossible to implement internationally. For such a treaty to come into effect, it would have to be negotiated, signed and ratified by at least the five Arctic Ocean states or the eight Arctic Circle states if it were to be effective. Thus all five or eight states would have to see the problem as severe enough to warrant the kind of international obligation and diplomatic commitment it would take to bring such a treaty into existence, and their governments would have to be able to approve of the resultant document to ratify it. Such negotiations and ratification processes take years even with full commitment. Moreover, notwithstanding their relative cooperation in the Arctic, these states are competing powers with competing interests, and in some cases they have ongoing territorial disputes with their Arctic neighbors that would have to be settled as part of the process. As a representative from an environmental law organization in the United States stated, 'Other people have the notion that there should be a framework that is similar to the Antarctic Treaty's, but ... I don't think that anybody is really talking about that seriously.'

Those in government and state-based organizations have also reiterated that position, which seems to imply that an Arctic treaty would be a hard sell. A government official in Norway was explicit on the matter: 'This [treaty] is nothing we are promoting', he said. 'We don't really see the need for any further legally binding newly written law.' Several respondents pointed to a flaw in the analogy with Antarctica: unlike Antarctica, the Arctic is an ocean and therefore the areas of the Arctic that are outside state boundaries are already governed by a widely accepted, binding international law: the United Nations Convention on the Law of the Sea. Hence, as was reaffirmed by the Arctic Five when they issued the 2008 Ilulissat Declaration, no additional legal instrument is needed. This perspective was reiterated by a representative of the Arctic Council, based on her personal experiences with the organization:

> The Arctic Ocean coastal states have made it very clear that there is no need for an Arctic treaty. We have the Convention of the Law of the Sea and build on that, to build additional agreements to regulate whatever needs to be regulated.... It seems like [the treaty] idea is fading.

Yet the fact that the Arctic, in contrast with Antarctica, is primarily a maritime region is not the only barrier preventing implementation of an Antarctic-type regime. There are also key differences in the human activities that occur in these two polar regions. The presence of indigenous groups, non-indigenous communities and many valuable resources set the Arctic apart from its southern polar counterpart. The Arctic has four million residents in dozens of communities; at most, the Antarctic claims just over 4,000, all scientists working out of research stations. Recognizing these key differences, some ENGOs have distanced themselves from the Antarctic model. As a Norwegian representative from an international ENGO explained, although 'lobby groups were ... proposing to make management systems − "rights"-systems − for the Arctic based on the Antarctic', his group did not support this agenda: 'The Antarctic is a lot of ice, a lot of nature, and some researchers, and a bit of very new fisheries, whereas the Arctic is a different story altogether, with a lot of people and lots of fisheries'.

States are not altogether resistant to treaties regarding the Arctic, and indeed some of these treaties are of great benefit to states. Bilateral treaties resolving boundary disputes, such as the one agreed to in 2010 between Norway and Russia in the Barents Sea, can make it easier for states to jointly manage waters and pursue the extraction of resources contained therein. Firmly set boundaries and regulations help attract companies that might not take the risk of undertaking expensive operations in unknown or disputed waters. But these treaties are firmly within the political control of the states that make them, and environmental protection is not typically the primary goal.

Protecting the People's Heritage:
The Indigenous Partnership Approach

Some international environmental groups bypass the treaty option entirely to focus on working with local populations. One such group is Pacific Environment, which, according to its website, 'provides direct support to communities and local organizations on both sides of the Arctic to strengthen coalitions, facilitate local community involvement in international processes, and to engage in direct advocacy'. Another, Ocean Conservancy, adopts a similar approach, calling for the various government management entities in the Arctic to control industrial drilling and fishing in the region, not just to protect the local fish and marine mammal populations for the sake of biodiversity but also to safeguard the subsistence lifestyles of the peoples of the Arctic.

A Norwegian representative of an international ENGO explained that her office is against an international treaty precisely because it would disadvantage the peoples who live in the Arctic region:

> This organization is very aware of the fact that there are a lot of people living in the Arctic . . . and they need to be heard. The idea of someone from outside coming in and making a treaty without being able to discuss it with people and making rules that don't fit with the people that live there, [this] is very much against the way we like to think how things should be governed.

Like the indigenous groups profiled in Chapter 6, who are wary of increased state involvement in the Arctic Council, her group fears that a state-based treaty would result in more state-based decisions, and these would exclude the people of the North, indigenous and non-indigenous, from any direct impact on the product of diplomatic negotiations. Under this scenario, northern peoples would have to content themselves with being represented by their state, which may or may not take their priorities into consideration, and this would negatively impact environmental, as well as indigenous, interests.

But even the many ENGOs who call for a treaty at the international level find themselves creating or attempting to create partnerships with indigenous groups to do work at the national or local levels. The choice to partner with northern indigenous groups is easily explained. Like ENGOs, indigenous groups often profess a concern for the health of the Arctic environment in which their members live and for the conservation of the wildlife that lives there as well. They also have a wide body of knowledge about the area that is valuable to environmental groups. Indigenous groups can work with ENGOs on wildlife protection issues that are of interest to both communities, monitoring where particular species congregate and migrate, whether their numbers have been rising or falling and whether there have been any underlying environmental patterns such as warming or melting that have been driving the changes they have seen. They also can work with ENGOs on public health projects – for instance, monitoring whether members of the indigenous group have been suffering adverse health effects

7.2   Idle No More demonstration, Anchorage, Alaska.

from persistent organic pollutants that have accumulated in the Arctic from their release elsewhere in the world, or whether they are being effected by heavy metal leaching from a nearby mine.

A representative of a Russian indigenous organization explained how many of his organization's concerns were explicitly *environmental*:

> We are concerned with pollution generally, but land pollution is our number one concern, pollution from mining and industrial wastes particularly from the Soviet legacy. Unfortunately, this is not all. . . . Diesel fuel and other heavy materials are extensively used in the Far North, and the negative impacts locally as well as climate change need to be better studied, and we share information with the Alaskan Aleuts.

Even though this organization's concerns are particular to its experiences in the Russian Arctic, there are enough similar concerns worldwide to encourage and facilitate cooperation among indigenous peoples on environmental pollution.

Indigenous peoples also have concerns about the welfare of terrestrial and marine wildlife in the Arctic, as many of them rely on hunting wild game and fishing for food. Anything that impacts the environment negatively would likely also harm the wildlife of the region, and that could have serious effects on Arctic inhabitants' access to food. It also implies that, from a survival standpoint, many native peoples must understand their environment and the impacts of any changes. As one indigenous Alaskan consultant pointed out,

> Alaska Native people consume anywhere between 400 to 2,000 pounds of wild food per capita. . .and that's probably more wild food than any populace in the United States, native or non-native. . . . And that means we still have an intimate connection to the environment and very profound knowledge and understanding about what is going on within ecological systems, species dynamics, you name it, weather, climate.

This knowledge can be a tremendous resource for environmental groups, and the concerns expressed above are concerns that the two groups share.

However, this does not mean that such a partnership is guaranteed to go smoothly. Just as there are many points over which indigenous groups and ENGOs agree, so, too, there are significant differences. One major sticking point that often derails potential partnerships is the desire of many ENGOs to protect certain species of wildlife, such as seals or whales, which some indigenous groups hunt as part of their traditional culture and look to as a source of income. Some environmental groups have made very clear their objection to the hunting of these animals, particularly for the purpose of earning an income.

It is important to note, however, that these difficulties differ from state to state and from the points of view of the different indigenous groups as well as the different ENGOs. Indigenous peoples do not all believe the same things or behave the same ways, and one group does not necessarily speak for the others (compare, for instance, the different political visions for the region's future held by the Inuit Circumpolar Council and the majority-Inuit Government of Greenland, discussed in Chapters 4 and 6). Even national affiliates of indigenous groups that are spread across states – the Inuit across the United States, Canada, Greenland and Russia, or the Saami across Norway, Sweden, Finland and Russia – may have different perspectives, as they respond to specific local conditions and state policies. One official of an ENGO based in Washington, DC noted this, explaining that his international ENGO had different relationships with various indigenous groups across the United States and Canada. 'Our relationship with the native community is on the whole pretty good in Alaska,' he said. 'If you go to Canada, it's a much more complicated situation.'

The Canada-based representative of the same international ENGO affirmed this, and noted that his organization has been specifically working hard to rehabilitate its image with the people of the North, specifically those in the Canadian Arctic Territory of Nunavut. This obviously includes indigenous groups, as Nunavut has a majority population of Inuit, 84 percent according to the 2011 Canadian Census. But this area also has a low employment rate – Statistics Canada reports that only 56 percent of Nunavut residents, and only 46 percent of Inuit Nunavut residents, were employed in 2011. The few employment opportunities that exist in the North, outside government employment, are in the resource-extraction industries, which are exactly the industries that the environmental movement generally wants to limit. When coupled with the general impression that ENGOs are insensitive to the needs and desires of local populations (e.g., through support for bans on some hunting practices), ENGOs are faced with a difficult task in their efforts to form community partnerships.

The Canadian-ENGO representative acknowledged this difficulty. He claimed that his group was trying to take into account the number of people in the region who wanted a better life for themselves and for their children. Nunavut has a relatively young population compared to the rest of Canada, with a third of the population under the age of 15 in 2011. Parents of these children want development, economic stability and a better life for their children, and the children want it for themselves. So in his view it is important for his group to acknowledge that development is a necessity in Nunavut, even as it seeks to protect the environment of the area from the potential devastation of many proposed development projects:

> We are shifting our language around economic development in the North. It's easy to say 'We shouldn't have mining in that environment' or 'We shouldn't have oil and gas development because it's so fragile.' It's easy . . . for us to say that, but they really don't have much else as far as to drive economic prosperity.

Because of this understanding that the (mostly indigenous) peoples of the region see resources as their main, and potentially only, driver of long-term economic growth, his organization is careful not to suggest that all resource extraction should be forbidden. To do so would set it at odds with the residents of the place that the organization is trying to preserve.

It would be wrong, however, to assume that all peoples of the North support development without reservations. Many northern peoples share the skepticism of ENGOs regarding the benefits of development and are wary of its environmental consequences. As the representative from a Russian indigenous organization pointed out, industrial development can hurt indigenous people too:

> Industry is developing the Arctic due to improved accessibility, and technological improvement is a driving force. But the pressures that come from industry, greater access to the North, and growing interest are interfering with subsistence lifestyles, including fishing and herding.

So it is by no means assured what position indigenous and other northern peoples are necessarily going to have on environmental protection. The circumstances in Canada are quite different from those in Russia; indeed, the representative of the Russian group above pointed out that his group's stance on industry and the environment is related to the environmental damage that occurred under the Soviet regime.

This all leads to an interesting point. Although ENGOs may adopt an international profile to address what they frequently identify as international problems, and, in some instances, they seek international treaties to address those problems, ultimately they are dependent on state action to accomplish their goals. Therefore, although international ENGOs may have global offices, like the Global Arctic Programme of the WWF, which sets the official policy for the ENGO, the policies and tactics themselves are decided upon and worked out within national offices in each relevant nation-state. That is to say, an international ENGO might have different policies in Denmark for Greenland than it has in Canada for Nunavut. And the relationship that an ENGO is able to strike with an indigenous group often depends on its history in that state, and what conditions the group faces there. The state still provides enough of an obstacle that an international ENGO's ability to interact with a particular community may not transcend state borders, even if both the indigenous group and the ENGO (and the environmental issue that it is addressing) do. Moreover, circumstances within the particular state, province, or community may dictate what the national office deems the most important goal to pursue at any given time, and this goal may not equate with the goals of either the other national offices or the international ENGO's approach or policy.

One example of this came from the US and Canadian representatives of one of the major international ENGOs. In 2010, The US office of this ENGO joined with the US government to support an effort to list the polar bear under the most threatened category – Appendix 1 – under CITES, the Convention on International Trade in Endangered Species of Wild Flora and Fauna. As an Appendix I species, i.e. one that is 'threatened with extinction', all commercial trade in polar bear products would be banned. The ENGO supported its effort with evidence from US-based science showing a decline in polar bears in Alaska. But the Canadian office of the same ENGO did not support the uplisting of the polar bear and the resultant ban on trade. The Canadian representative of the international ENGO had various reasons to offer for this discrepancy in the two national groups' policies, but foremost was that Canadian science did not indicate a decline in polar bear populations. Not mentioned, but also important, was that Canadian Inuit find employment opportunities leading polar bear hunts, an option closed to Alaska Natives due to the polar bear's listing under the US Endangered Species Act. On top of all this, a representative of the international ENGO's main office for the Arctic suggested that another reason to oppose the listing in Canada was a fear of driving the hunt underground, which could lead to a trade in products on the black market. The representative from the international office acknowledged that there were differences in policy among the group's national offices but asserted that this was inevitable given the decentralized nature of the organization and the fact that so much of its activities were oriented toward influencing national policies.

Communicating the Vision: Working with Others

Regardless of whether ENGOs work alone towards an international treaty or partner with indigenous groups, they must gain buy-in for their vision if they hope to achieve their goals. They have to cooperate with the national government and state/provincial governments, and they may even have to collaborate with the industry groups whose activities they generally work to limit. As a Greenlandic representative of the Inuit Circumpolar Council pointed out, his group was against the decision of the Greenlandic government to allow offshore drilling. But, he said, despite their disagreement, 'we can't deny our government's right to do that'.

In this case, an attempt to deny that right would question political authority with regard to the extractive industries. And, as was discussed earlier, there is a very strong drive to drill for oil in Greenland due to the region's extreme need for a revenue source in order to achieve the Greenlandic dream of independence. As a Danish official pointed out,

> If they [the Greenlandic government] had to choose between a lot of oil and a lot of money, and . . . then an oil disaster, they would probably take [the latter], because the alternative is being forever on social welfare from Denmark.

Under such circumstances, it is difficult for indigenous groups, ENGOs, or anyone else who seeks to limit economic activity in the interest of environmental preservation to expect to have much impact on state behavior.

Moreover, states pursue their own interests, and signing international legislation such as the treaty that some ENGOs would like to create for the Arctic would by its very nature reduce the ability of states to pursue unilaterally their own interests. States generally prefer to enter into international arrangements by their own choice and not through pressure, and on most occasions they would rather avoid them altogether. Furthermore, if a treaty modeled after the Antarctic Treaty System were to designate the Arctic as the 'common heritage of humankind' or some other commons to which global protections (or access) would be guaranteed, then the five Arctic Ocean coastal states might well lose some of the sovereign rights to EEZ and outer continental shelf resources that they have gained in the region through UNCLOS. This move could have dangerous consequences given that non-Arctic states like China and state organizations like the European Union are already showing interest in the region.

If ENGOs have a hard case to make when lobbying the state, their ability to work with industry groups is even more hindered. Their relationships with industry are generally described as strained at best and outright hostile at worst, again depending on the state/territory, the ENGO and the industry in question. In response to environmentalist challenges, major oil companies point out that they are well aware of the potential for environmental damage

7.3   Greenpeace attempts to block passage of the Finnish icebreaker *Nordica* on its way to assist Shell's exploratory efforts in the Arctic, near Fehmarn Island, Germany.

caused by their activities, especially in the wake of the 2010 Deepwater Horizon incident in the Gulf of Mexico. They are also aware that drilling in the North, in icy or ice-covered waters, poses particular risks. ENGOs and the oil companies differ, however, in the degree to which they find the risk acceptable. The oil companies take the perspective that risks are always present but can be managed or minimized. A representative from one major international oil company asserted her company's commitment to the environment, particularly with regard to offshore drilling: 'We believe that we should have absolutely no impact on the environment.... And whether we work in shallow water or deep water, that is the same. You just work and identify the risks.'

But ENGOs believe that risk cannot be appropriately minimized in the Arctic, and that if drilling commences (or, in some regions, continues) a spill will inevitably ruin the environment. A Washington, DC-based official with an ENGO stressed that no matter how great or small the risk is, there will someday be an accident. That would be catastrophic, because his group '[does] not think that there is a capacity to respond to a spill in the Arctic. There will never be the capacity.' There is little room for reconciliation between these beliefs: either a risk is potentially manageable or it is inherently unmanageable. Thus, there appears to be an insurmountable gap between these two opposing viewpoints and the organizations that espouse them.

Moreover, some in the resource-extraction business believe that even if they did sit down to negotiate with ENGOs they would not be able to do enough to please them with regards to environmental safeguards and protections. Several industry representatives whom we interviewed (particularly in the United States) voiced their opinion that the central objective of ENGOs ultimately is not to protect the environment but to prevent economic activity. Thus they identified an unbridgeable chasm between themselves and the ENGOs, since a corporation's reason for being is to engage in economic activity. One official from an Alaskan extractive industry association, in particular, felt that ENGOs would not be happy until his industry ceased to exist, and that many of the arguments that ENGOs make in favor of increased regulation are simply a means toward their end-goal of making it harder or impossible for companies in his industry to do business. Turning to the high-profile oil and gas industry, he asserted,

> They don't believe that oil and gas activity should be happening out in the Arctic Ocean. We of course vehemently disagree . . . . So that relationship is extremely strained at the moment and will continue to be so.

An Alaskan elected official seconded this point of view: 'We believe that in Alaska we're often set between people who support sustainable development and people who support no development.' Of course, most if not all ENGOs vigorously dispute this characterization of their agenda, claiming that they only wish to ensure that development occurs in a sustainable and environmentally safe way. But it would be extremely difficult for ENGOs to remove these differences in perception in order to work with oil, gas, or mining corporations.

An Usullied Future?

Although the vision of an Arctic that is untouched wilderness, a place of nature and animals with stringent limits on human pollution and degradation, undoubtedly exists, the Arctic is actually a complex region of human life, economics, governments and environments. ENGOs in many countries are working to help improve the lives of individuals in the North, and to promote a vision for the Arctic that involves both environmental protection and human development.

Their success in promoting this vision depends often on their circumstances, which in turn frequently depend on the political and social environments in which they operate. International ENGOs have more freedom to propose nature-centric policies, such as elaborated in the WWF

Global Arctic Programme's proposal for an international treaty to protect the Arctic marine environment. National and local affiliates, by contrast, tend to be more pragmatic, balancing what they would like to have with what they think they can get, while recognizing, from their on-the-ground work, the need to build coalitions. National offices also face different challenges and opportunities when it comes to working with residents of the North, indigenous and otherwise. Some indigenous groups are more concerned about the environment, or certain aspects of it, for a range of reasons. Some peoples of the North are more concerned about ENGOs because their global initiatives may unfavorably impact their income. There is no monolithic group that speaks for all of the nearly four million residents of the Arctic, just as no one ENGO can be held up as the exemplar of all the rest.

ENGOS, indigenous groups, corporate interests and states are projecting their visions for how the Arctic's remaining wilderness areas will be accessed. The changing contours of ocean, ice and land, coupled with the region's complex and evolving cultural and political landscapes, make the Arctic unique, even if many of the challenges and dynamics faced by Arctic actors parallel those found in other parts of the world. Questions concerning different types of development, human and non-human rights and conservation-versus-preservation are still being met with competing answers in the Arctic, informed and challenged by the various imaginaries explored in this book. As the future of the Arctic is established, the influence of the ENGOs will be linked to their ability to negotiate across these visions for the region.

8.1 Movie poster for *Ice Station Zebra*.

CHAPTER 8

NORMALIZING THE NORTH

With the emphasis on the overlapping, mutually reinforcing, and at times competing Arctic imaginaries that have jostled for attention in the last couple of decades, it is possible to forget yet another imaginary that had to be displaced before the ones discussed in this book could come to their fuller expression, and that predictably re-emerges during periods of renewed East–West tensions. This is the imaginary of the Arctic as a geopolitical chessboard.

In the aftermath of World War II formal hostilities, big-budget Hollywood filmmakers reflected and contributed to the popularization of a chilling imaginary: that of a faraway Arctic made forebodingly nearer by the uncertain nuclear standoff between the world's two superpowers. As the Cold War intensified, the Arctic became the site of more nuclear weapons than practically any other part of the world, and the 1952 film *Arctic Flight*, set in the Bering Strait, mirrored growing American fears of Soviet espionage that were peaking during the McCarthy era. A decade later, *Fail-Safe* (1964) raised the ante from espionage to possible superpower nuclear conflict as filmgoers watched a flight of US Air Force Vindicator bombers, on patrol over the Arctic Circle, receive the dreaded message: 'Attack Moscow'. In *The Bedford Incident* (1965), the Arctic again was the stage for conflict, this time rooted in public anxieties surrounding nuclear confrontation. In this film, newspaper reporter Ben Munceford was embedded on a US Navy destroyer, where he observed, helplessly, the Ahab-like American captain unrelentingly track a Soviet submarine between Greenland and Iceland, as he aimed to transform the Cold War into a 'hot' one. *Ice Station Zebra*, released in 1968 – the year Soviet troops invaded Czechoslovakia – brought together American and Soviet military, spies and search-and-rescue efforts in a taut race to reach the Arctic, culminating in a fire fight and death on an Arctic ice floe. Then during the 'Second Cold War', marked by the United States' boycott of the 1980 Olympic Games in protest of the Soviet Union's invasion of Afghanistan, the alarmingly titled *World War III* (1982) again exploited popular fears about vulnerable northern borders, as the Arctic became a site for Soviet invasion and sabotage of the Alaska Pipeline.

Like the imaginary of Arctic integration promoted by Vilhjalmur Stefansson, Mikhail Gorbachev and Sarah Palin, discussed in Chapter 1, the Arctic in these films was portrayed as a region of proximate interaction among circumpolar territories. But, *contra* Stefansson (and Gorbachev and Palin), the Arctic imagined by Cold War Hollywood producers and military strategists was anything but 'friendly'. Indeed, the region was prized by the two superpowers for its strategic utility as the shortest route for projecting missiles or launching an invasion between the two countries. As such, the Arctic region was largely left in a frozen state of stalemate and military stratagems.

The first thaw in the Arctic's icy standoff began in 1985, when Mikhail Gorbachev became General Secretary of the Central Committee of the Communist Party of the Soviet Union. Gorbachev proposed a normalization in Arctic relations through the elimination of nuclear weapons in northern Europe, a limit to naval operations in the maritime Arctic and multilateral, trans-border collaboration on scientific research, environmental protection, marine transportation and indigenous peoples' affairs.

Gorbachev's proposal to dissolve 40 years of Cold War contestation across the Arctic was met, not surprisingly, with suspicion by the other northern and western nations. But then, on a November 1989 visit to St Petersburg (formerly Leningrad), as the first hammer strikes to the Berlin Wall were falling, Canadian Prime Minister Brian Mulroney significantly asked: 'And why not a council of Arctic countries eventually coming into existence to coordinate and promote cooperation among them?' With the Cold War thus fading in the 1990s, the beginnings of an institutional structure that could facilitate circumpolar scientific research and Arctic-specific agreements began to take shape among the eight states with territorial sovereignty north of the Arctic Circle. A new notion of security thus began to be articulated in the Arctic: 'Human security'. While the concept was left vague, it basically presented a vision that, although incorporating state-centric imaginaries, nonetheless could de-politicize and limit international tensions.

In addition to a new vision of northern cooperation, the end of the Cold War also opened the region to new imaginaries of territorial enclosure, Arctic indigenous peoples' self-determination, environmental protection, and resource access and control. At some level, these imaginaries had always been present but in the previous decades they had been complicated by tense superpower relations. Yet it was not just the warming of international relations that fostered conditions for these imaginaries to become more viable (and contested). Interest in the region was also stoked by the advent of global climate change and the understanding that the Arctic was likely to be a very different physical place in the near future. In light of the challenges and opportunities that a warming Arctic seemed to promise, imaginaries that had always been latent began to crystallize and make their way into the discourse of politicians, the corporate and civil sectors and, perhaps most importantly, mass media.

With all these imaginaries swirling, the 2007 Russian flag-planting was greeted with a media uproar, centered on speculation that the Arctic could again become a site of intense conflict. The underlying question in this media build-up was, of course, the issue of sovereignty: Who ultimately had control of, and responsibility for, the Arctic? It is this assemblage of events and discourses that then led to a rather instinctive move by state actors: to assert that the state – and the system of territorial sovereignty that sanctifies it – was in control, that this control was established and grounded in law, and that there really was no ambiguity about states' jurisdiction over the Arctic. This message was proclaimed loudly and clearly in the Ilulissat Declaration of 2008.

Yet, as was explained at the beginning of this book, the Ilulissat Declaration, as self-evident as it appeared to the five states that signed it, was criticized for being too restrictive and state-centered because only the Arctic Five were involved. The backlash against Ilulissat suggested that successful efforts to govern the Arctic would need to take a less exclusionary approach. Yet, the critical reaction to Ilulissat did not mean that the state system was being fundamentally challenged in the Arctic. Perhaps the signatories to the Ilulissat Declaration had exceeded their authority, but the process of establishing a state-dominated order in the Arctic was already well under way.

The UNCLOS agreement and the subsequent establishment of the Commission on the Limits of the Continental Shelf were clear steps in the direction of establishing sovereign rights to significant and possibly contested parts of the Arctic. In addition to this international legal framework, the Arctic Council – a high-level intergovernmental forum – has also proven adept at providing an open environment for discussion of diverse circumpolar issues, including jurisdiction over the still quite extensive Arctic marine areas that remain outside sovereign territorial control. This is not to say that the Arctic is a settled place with regard to governance. As this book has shown, the Arctic remains an area of various and contesting imaginaries, each of which is unevenly contributing to the future of the region. Rather, we stress that there is in fact a process in place through which these various imaginaries are tamed, made compliant and finally operationalized through applications and adaptations of the principles of sovereign rights and international law.

As each of the previous chapters has discussed in detail, the Arctic is undeniably a dynamic and evolving region that consists of the *interaction* of a variety of imaginaries rather than a place of conflict among 'pure' visions for the future. The 'conflict' so readily depicted by Hollywood, the media, and at times even by Arctic stakeholders themselves, is an uneven process of contestation, compromise and normalization as states and non-state actors create, implement, or respond to changes in the Arctic from specific, but overlapping, imaginaries. The Arctic Council is the institutional space that not only mirrors these contesting imaginaries but that actively works to bring these imaginaries into more constrained reflections of their 'purer' forms. The aim of this process is, in the end, to increasingly normalize the Arctic and make it like any other place: a region that is governed by states, in which law and order is maintained so as to facilitate investment and commerce without major conflict.

Normalizing the (Seemingly) Abnormal

The Arctic Council's efforts at 'normalizing the North' have been characterized less by unbending application of international law than by the development and deployment of creative adaptation strategies. Rather than shutting out the alternative imaginaries depicted in Chapters 2 through 7 of this book, the Arctic Council has worked hard to co-opt them. In the process, the statist norms asserted in the Ilulissat Declaration have been affirmed, but they have also been creatively adjusted in order to account for the fact that, in some fundamental ways, the Arctic *is* different.

Perhaps this is most clearly evidenced in the ways in which the Arctic Council has coped with the problems of organizing search-and-rescue operations in the North. In one sense, there is little remarkable about the Arctic Council's Agreement on Cooperation on Aeronautical and Maritime Search and Rescue in the Arctic (the Search and Rescue, or SAR, Agreement), beyond the point frequently noted by commentators that it was the first legally binding treaty negotiated under the Arctic Council's auspices. Prior to negotiation of the Arctic Council SAR Agreement, several international legal instruments had already been agreed to that specify the responsibilities and obligations of coastal states in emergency situations, including Article 98 of UNCLOS and the much more detailed International Convention for the Safety of Life at Sea (SOLAS). In addition, a convention specific to search-and-rescue operations – the International Convention on Maritime Search and Rescue – establishes minimum levels of infrastructure that coastal states must have available for search-and-rescue operations and encourages the sharing of resources among regional coastal states. This convention is particularly relevant to the Arctic Council SAR Agreement because it supports regional seas conventions under which member states make their own agreements for allocating and sharing resources. Thus, to a large extent the Arctic Council, in facilitating the SAR Agreement, did little but apply to the Arctic what had already emerged as a global norm.

However, unlike other regions' agreements, the Arctic Council's SAR Agreement covers all water *and land* north of the Arctic Circle (and, in some cases, south of the Arctic Circle as well). As was discussed in Chapter 2, the sectoral wedges delimited by the Arctic Council agreement resonate with a perspective in which the Arctic exists outside the land-water divide that underpins the modern state system, and it (inadvertently) suggests an imaginary in which Arctic space – whether land or water – exists as a *terra nullius* that potentially can be claimed by any state bold enough to assert its presence there.

In fact, as was also discussed in Chapter 2, the SAR Agreement is explicit in distancing itself from this *terra nullius* imaginary. But the agreement *does* borrow from the somewhat less extreme 'frozen ocean' imaginary discussed in Chapter 3. According to this imaginary, although the Arctic is not a region in which the norms that define and divide categories of Earth's surface are universally transcended, it *is* a region in which they must be adjusted to account for its exceptional nature. In fact, the Arctic Council agreement is explicit on this point, noting that the treaty is needed, in part, '[because] of the challenges posed by harsh Arctic conditions on search-and-rescue operations and the vital importance of providing rapid assistance to persons in distress in such conditions'. In other words, it matters that the Arctic is a frozen (or, strictly speaking, partially frozen) ocean, notwithstanding the premise of international law that its physical state is irrelevant.

By emphasizing the Arctic's geophysical uniqueness, the Arctic Council SAR Agreement is in some ways designating the Arctic as an *exceptional* region. And yet, by resorting to a sectoral division of the Arctic Ocean that efficiently allots responsibility to individual states, the agreement manages to divert any call for a more comprehensive, non-state-based management regime. Rather, the agreement places the Arctic within the overarching, and 'normal', political framework of sovereign statehood. The agreement thus reaffirms that the Arctic is both in the domain of *individual states* (as each of the eight signatory states has responsibility over a specific portion of the Arctic) and the *community of states* (as the eight states have committed to work together to facilitate the sharing of resources). In this sense, the agreement is a metaphor for the actions of the Arctic Council as a whole: through creative interpretations of existing norms – including, at times, interpretations that draw on alternative Arctic imaginaries – the Arctic Council successfully reproduces the norms of the state system and the rule of law, thereby facilitating the intensification of the region as a zone of commercial activity.

Controlling the 'Nature Reserve' Imaginary

Just as the Arctic Council's Search and Rescue Agreement has selectively incorporated elements of the 'frozen ocean' imaginary, its scientific working groups have incorporated (and thereby partially neutralized) elements of the 'nature reserve' imaginary and the environmental concerns that motivate environmental non-governmental organizations (ENGOs) to attempt to influence Arctic policy. As was discussed in Chapter 7, ENGOs have devoted increased attention to the Arctic since the 1990s, driven by the region's perceived exoticism, its charismatic megafauna, its ecological vulnerability and, most recently, its position as a dramatic indicator of global climate change. As Greenpeace's website explains, 'Inevitable mistakes would shatter the fragile Arctic environment.' Yet, notwithstanding the Cold War concerns depicted at the beginning of this chapter, the 'Arctic' – and certainly the Arctic's environment – has until recently occupied a peripheral position in the southern capitals of states with Arctic territories. Long seen by policy makers as just an extension of existing political, economic and social systems, the Arctic – never legally defined nor cartographically clear – has only recently emerged in some regards as a distinctive 'region', albeit a shifting and unstable one.

The scientific research carried out by the Arctic Council and made widely available through print and electronic distribution has been instrumental in placing the Arctic's marine and terrestrial regions on the 'front lines' when it comes to experiencing and understanding the consequences of global environmental changes. Thus, in addition to the adjectives 'pure', 'pristine' and 'frozen', new words are increasingly appearing in the lexicon of Arctic adjectives: 'polluted' and 'changing'. The Arctic environment, reinterpreted as threatened by industrial pollutants and climate-changing processes originating from outside the region, has offered the environmental community an imaginary around which to rally.

As was discussed in Chapter 7, environmental organizations clearly recognize that the Arctic cannot somehow be removed from global political and economic processes to form a pure nature reserve separate from humans, and they are aware that in any event such a solution would be completely unacceptable to the Arctic Council's member states and permanent participants. Nonetheless, the concerns raised by the environmental community are profound and have commanded attention. The image portrayed by ENGOs of Earth's last pristine place becoming blackened by the encroachment of development, economic growth and industrial pollution is powerful and provides an influential counterweight to state and corporate visions of a region ready for resource exploitation (i.e., the 'resource frontier' imaginary discussed in Chapter 5).

For their part, the Arctic Council's eight member states recognized early on that environmental issues could form a benign platform around which to generate multilateral cooperation and to build trust in the council's competency as the definitive regional entity able to handle matters concerning the Arctic. Environmental concerns, such as the melting of the Arctic Ocean's ice cover, as well as pollution and the protection of living resources, stand out as issues that transcend state borders and that require multilateral responses. A shared understanding that there is a fundamental mismatch between territorial and ecological boundaries is something around which the Arctic Council and the international environmental community have been able to find common cause. Recognizing that it would be excessively costly for each Arctic state to sponsor its own comprehensive Arctic research program, and also that the research conducted by national programs could be limited by jurisdictional boundaries, the eight Arctic states have promoted the Arctic Council as the main body for interstate Arctic cooperative scientific research. By actively and consistently seeking input from hundreds of leading Arctic researchers, indigenous peoples' representatives and experts from nations outside the Arctic, the Arctic Council has been delivering a clear message to the international environmental community that through the sponsorship and coordination of scientific research it will make a major contribution to shaping the Arctic's future.

To this end, the Arctic Council's six working groups engage in issues such as monitoring, assessing and preventing pollution in the Arctic, climate change, biodiversity conservation and sustainable use, emergency preparedness and prevention, and the common concerns and challenges faced by Arctic residents.[1] The reports produced by these working groups provide influential scientific assessments and technical recommendations, and they form the basis for policy suggestions not only to the Arctic Council's broader membership but also to the environmental community. The working groups' research serves to communicate that the Arctic Council places information collection on environmental issues among its top priorities, but it also demonstrates the Arctic Council's member states' collective commitment to exercising environmental leadership throughout the Arctic.

At the same time, however, scientific assessments, while endeavoring to achieve objectivity, nonetheless involve choices and priorities. Thus the reports of the working groups must be understood not simply as presentations of information but as part of a deliberate process to produce and frame knowledge. In this manner, the research and policy recommendations offered by the Arctic Council's working groups re-frame many of the political ramifications associated with climate change as technocratic problems that are addressable through technical means, thereby obviating the need for more drastic solutions, such as setting aside the Arctic as a nature reserve or curtailing the rate of industrial growth elsewhere in the world. The implication is that, through careful and collective application of region-specific policies, which are derived from comprehensive scientific research that indeed accounts for the concerns of the environmental community, the issues that are negatively impacting the Arctic can be addressed. In this manner, the Arctic Council, led by its eight member states, but including the full participation of its permanent participants and governmental and non-governmental observers, has attempted to attune its collective expertise to the managerial organization of the Arctic for the purposes of 'sustainable development' in a peaceful and cooperative manner. Thus, although effecting remarkable environmental leadership in the Arctic and providing a high-level international forum in which the concerns of the environmental community may be heard, the Arctic Council's goals are limited in that it is not seeking to lead a broad international effort to curtail the industrial processes that are negatively affecting the Arctic's climate.

A case in point is the Arctic Council's Arctic Climate Impact Assessment (ACIA) and its policy recommendations. This groundbreaking assessment, initiated during the US chairmanship of the Arctic Council (2000–2) and released to the public during Iceland's chairmanship (2002–4), was the first comprehensive regional assessment of climate impacts. This circumpolar research project collected, evaluated and synthesized knowledge on the processes and consequences of climate variability and change for the entire Arctic region. The ACIA resulted from the work of more than 300 leading Arctic researchers, indigenous representatives and other experts from 15 nations. According to its website, 'The ACIA is the world's most comprehensive and detailed regional climatic and ultraviolet radiation assessment to date and documents impacts that are already felt throughout the Arctic region.'

The ACIA was largely responsible for shifting the commonly held perspective of a *frozen* Arctic to that of a *changing* Arctic, and it helped focus the Arctic Council's other working groups toward producing specific scientific assessments of the environmental and economic consequences of climate change in the region. Although broadly supporting the dominant perspective in the climate science community – that efforts are needed to mitigate the risks of climate change – the ACIA supported another important dimension to policy recommendations because, as it straightforwardly expressed,

Climate change is inevitable, indicating that continued adaptation is needed. Adaptation to climate change and its impacts in the Arctic must take into account the especially sensitive and vulnerable natural and human systems of the region. Special attention needs to be paid to strengthening the adaptive capacities of Arctic residents.

The shift in attention to *adaptation* displays a significant political perspective. While 'special attention' gets placed on the necessity of Arctic residents to strengthen their 'adaptive capacities', little is said about holding distant polluters answerable for their impact on Arctic peoples and environments.

The ACIA, while obtaining broad input and representing multiple perspectives, including contributions from the environmental community itself, limits the universe of possible solutions so as to avoid potential contradictions with the Arctic Council's member states' own particular domestic priorities regarding climate change and environmental protection. The ACIA illustrates the Arctic Council's effectiveness at bringing the broad-based environmental community (and also indigenous peoples and scientific expertise as well as interests from outside the region) into the discussion, giving them 'seats at the table' and incorporating their knowledge and perspectives into policy recommendations. But it also illustrates how such multilateral cooperation will be pursued, maintained and supported only so long as it works within, and does not challenge, state interests.

Indigenous Imaginaries

Unlike the Antarctic, whose legal regime evolved to suspend territorial claims, stop possible militarization and preserve a seemingly unspoiled environment for scientific research, the Arctic's emerging governance structure includes the development needs of northern residents who, in the case of the region's indigenous peoples, have been living in the circumpolar North for millennia. In the Antarctic, application of the 'nature reserve' imaginary has in some ways been achieved, but such an imaginary in the populated Arctic is practically impossible, at least in anything approaching its 'pure' form.

A successful effort to create a normalized space of predictable governance in the Arctic therefore must account for the interests of its most longstanding residents. Even before the latest surge of enthusiasm for Arctic resource development, individual Arctic states had realized the need to work with their native northern peoples in order to facilitate governance, expand settlements and establish a business-friendly environment for investors from outside the North. This led to significant legislation, particularly in the United States, Canada and Greenland, that codified native rights and recognized land claims and autonomy arrangements. Yet, even as this was occurring, indigenous peoples' organizations, and particularly the Inuit Circumpolar Council, maintained a strong commitment to a transnational identity and the need to be politically organized on an international level as well.

From the perspective of these groups, a transnational focus was natural. For one, many of the indigenous peoples of the Arctic are *de facto* transnational, since, as groups located in more than one state, they share interests and challenges that are not confined to one particular country. In discussing and advocating particular issues, members of transnational indigenous groups have found building links across state borders to be strategically beneficial. Many of the greatest concerns facing circumpolar indigenous groups are not confinable to the state level – seal skin bans, for example, are enacted by multiple states; air and water pollution is traced to areas outside individual state boundaries – and therefore indigenous groups benefit by bringing these issues to the international stage. Because opportunities for influencing political goals may be limited within their own states, and resolution of many of these issues requires international cooperation, indigenous groups seek access to international declaration- and treaty-making processes.

As was discussed in Chapter 6, indigenous groups of the North have also long struggled with the concept and the enactment of state sovereignty that has been imposed on their traditional lands and waters. As was indicated in the ICC Declaration on Sovereignty, although the power of governments in the south cannot simply be denied, it can nonetheless be problematized. States' claims to absolute authority over distant stretches of the Arctic, of which they often have little knowledge and limited presence, can be cast as spurious. Furthermore, the assertion of a presence, particularly a historical one, relies directly on indigenous peoples themselves. As such, indigenous peoples have demanded that any practice of sovereignty in the Arctic must integrally incorporate the native peoples that actually inhabit these areas. From this perspective, a push to transnationalize Arctic governance strikes a chord with indigenous groups because such governance at least holds out the hope for indigenous peoples to be included in important policy shaping, and ultimately policy making, in the Arctic.

It was in this context that the largest transnational indigenous group, the ICC, emerged as a strong advocate and in fact an ally when the Canadian government endorsed the creation of the Arctic Council. Subsequently, a key aspect of the Arctic Council's multilateral cooperation has been the inclusion of six regional indigenous peoples' organizations as permanent participants, and it is telling that five of the six permanent participants represent cross-border constituencies.[2] These groups are concerned with, among other issues, promoting the rights and interests of indigenous peoples throughout the circumpolar North, within a state's territorial borders but also quite frequently independent of those lines.

The position of indigenous groups in the Arctic Council is certainly innovative in the world of intergovernmental affairs, and it is often pointed to as an example of successful participatory governance. It is, furthermore, something that indigenous groups want to protect, being wary, for instance, of allowing new permanent observers to join or, as in the case of the Ilulissat Declaration, becoming marginalized by northern states that consider themselves the most legitimate players in the Arctic. Full inclusion as permanent participants on the Arctic Council reinforces indigenous peoples' standing in international law and helps corroborate self-government processes. But at the same time, opinions vary among indigenous peoples' representatives regarding their actual role on the Arctic Council itself. For example, an indigenous activist whom we interviewed found the role played by the six permanent participants on the Council practically irrelevant:

> All of the indigenous peoples are relegated to advisory posts as NGOs, they don't have any meaningful voice or any say about anything. And of course the Arctic Council can say, 'Oh yes, we have consulted with indigenous people,' but that is meaningless. It basically means we [member states] let the indigenous people say what they are going to say and then we do whatever we want to do anyway.

The ambiguity of indigenous people's placement as both inside and outside state borders and policies is something that the Arctic Council must negotiate, a process that is further complicated by the council's multilateral participatory, though state-led, structure. Much of the work done by the Arctic Council, while certainly including indigenous peoples' participation and providing a forum to voice many of their priorities, still at times delimits indigenous peoples as 'traditional' and their geographies as 'local'. Indeed, the emphasis placed by outside observers on the traditional relations between indigenous peoples and their environments can itself be derivative of an imaginary that may serve to restrict indigenous peoples to pre-conceived political and spatial locations. As expressed by a consultant to indigenous peoples' organizations, 'The aboriginals just don't have the capacity or the might' to draw international negotiations to their perspectives. This demarcation of the roles played by permanent participants thus serves to maintain state-centric political-economic relations within the Arctic Council, consequently limiting indigenous enactments of nationhood.

As was discussed in Chapter 4, the myth of the traditional native is double-edged: on the one hand conferring power, yet on the other establishing a clear limitation in terms of agendas and alliances that can be pursued. For instance, by focusing on indigenous peoples' 'local' and 'traditional' concerns, states – and hence also the Arctic Council – are able to diffuse more problematic political claims by indigenous peoples that seek to step beyond local constraints (e.g., the pursuit of independence). In this way, the inclusion of indigenous groups in the Arctic Council helps to legitimize, while at the same time tempering, what is fundamentally a state-centered Arctic. Participating on the Arctic Council is thus, in certain ways, not unlike native land claims agreements whereby indigenous/state relations are normalized: state sovereignty remains unthreatened and even supported by the presence of indigenous peoples, while indigenous peoples maintain a unique position with regard to traditional lands and waters.

This means of incorporating indigenous peoples into Arctic governance gives the permanent participants an unprecedented voice on the international stage, as expressed to us by an activist with one of the less powerful permanent participants: 'The Arctic Council is the single individual organization in which natives participate as full participants. This presents us a democratic stage to allow full, equal position and involvement in policy.' Yet, for groups seeking greater political self-determination – for instance the Greenlandic political elite – the Arctic Council has limited relevance. Supporters of Greenland's bid for statehood reject the constraints that are often associated with the concept of indigeneity and the clashes and misperceptions it engenders between those who hold different presuppositions about how the world works. Greenland therefore does not use its position in the Arctic Council in any way to pursue its sovereignty ambitions. Ultimately, participation in the Arctic Council as anything less than a sovereign state is seen by the Greenlandic elite as incompatible with their state-seeking ambitions. This was demonstrated at the May 2013 Arctic Council Ministerial Meeting in Kiruna, Sweden, at which representatives of the Government of Greenland chose to absent themselves rather than be seated behind the Danish foreign minister.

With statist norms still providing the dominant vision for the Arctic Council's activities, challenges to the existing order of sovereign states have no place in the Arctic Council, whether they are from a group that questions the legitimacy of sovereignty as an institution or from a group seeking its own status as a sovereign state. Although Greenlandic statehood would, in one sense, reaffirm the state system by asserting the pre-eminence of territorial sovereignty as the fundamental global organizing principle, it would also undermine it by suggesting that any individual state's existence is subject to negotiation and therefore not 'natural'. For these reasons, the indigenous statehood imaginary is a non-starter within the council. Indeed, the power that indigenous permanent participants have been accorded by the Arctic Council's member states diffuses such aspirations.

In the case of Greenland, therefore, the self-rule government's ambitions are simply treated as a Danish domestic issue, mirroring a common theme within the institution of sovereignty wherein there is a strict (if idealized) separation between 'domestic' and 'international' affairs. If Denmark chooses no longer to permit Greenlanders' participation in their delegation, then the Arctic Council would surely respect that decision. If Denmark chooses to allow full independence to Greenland, then that decision would be accepted as well. In this way the Arctic Council diffuses the Greenlandic imaginary of statehood by ignoring it. As with the Search and Rescue Agreement and the working groups' scientific work, adaptive measures are taken by the Arctic Council to normalize the region. Whether through the inclusion of shared regional concerns within a discourse of indigeneity or through delimiting indigenous political issues as 'domestic', the Arctic Council manages to stay the course in extending and solidifying intergovernmental cooperation in a region that increasingly is rooted in the institution of state sovereignty.

The Arctic Community

While the Arctic Council likely will remain removed from any future negotiations regarding Greenland's representation, there is a thornier issue that dominated news stories about the council during 2012 and 2013, and that likely will remain a topic of concern for years to come: the question of which countries should be granted permanent observer status. As the Arctic Council emerges as the single most influential intergovernmental Arctic institution, it is increasingly being approached by non-Arctic states with their own Arctic priorities and justifications for inclusion in the region's governance. Attracted by the imaginary of the Arctic as a 'resource frontier', states from outside the region are looking to be included in Arctic policy and decision making, and, with the Arctic Council evolving into a policy-making as well as research body, it is not surprising that these countries have focused their sights on the Arctic Council. This sets the stage for a true internationalization of the region, but it also – like the other challenges considered in this chapter – presents prospects for the reproduction and maintenance of the dominant, statist governance structure.

To provide a forum for state and non-state parties from outside the region, the Arctic Council has developed a role for observers. Observers can submit written statements and participate in working groups, and they can also address ministerial meetings if invited to do so by a full member. From the Arctic Council's inception through 2009, six countries (France, Germany, the Netherlands, Poland, Spain and the United Kingdom) as well as nine intergovernmental organizations and 11 non-governmental organizations were granted permanent observer status. Other states have been frequent ad hoc observers, but to be admitted as an ad hoc observer one must reapply for each meeting, and since the decision to admit an observer must be unanimous entities seeking ad hoc observer status are continually at risk of rejection.

Throughout the Arctic Council's first decade, ad hoc observers were routinely approved if they sought permanent observer status. But as the number and geographical diversity of requests, and interest in the Arctic, increased, the practice of approving permanent observer status was suspended at the 2009 Arctic Council Ministerial Meeting in Tromsø, Norway, pending further consideration by the council's membership. In general, the Nordic countries have been most supportive of expansion. They calculate that an enlarged council will allow them to extend their diplomatic reach while establishing a forum wherein Arctic and non-Arctic countries can work together to facilitate external investment in the region. The permanent participants have generally opposed expansion, for fear that their voices and concerns will get lost in a more crowded field of non-Arctic states and organizations. More generally, the permanent participants are concerned that if the Arctic were to be rescripted as a zone of *interested parties* rather than one of *territorially defined stakeholders* (represented institutionally by states and permanent participants), their potential influence could be diminished. Paralleling the general debate about Arctic Council enlargement is the more specific question of the EU's application for permanent observer status, which has been blocked by Canada (with the support of the Inuit Circumpolar Council) due to the EU's ban on seal and fur imports.

In fact, the significance of the ongoing debate over which entities should be granted permanent observer status may be little more than symbolic. Once an observer is admitted, its rights are identical, whether it is an ad hoc or permanent observer, and even permanent observers must have their status periodically reaffirmed. For these reasons, Norwegian defense expert Rolf Tamnes suggested in a January 2013 public address that the ongoing debate over expansion of the permanent observer category was 'much ado about nothing'. Significantly, though, Tamnes then noted that if any party felt that obtaining permanent observer status was important then it truly *was* important, because a spurned applicant might seek a different venue for airing its concerns.

The permanent observer issue came to a head in May 2013, at the Arctic Council Ministerial Meeting in Kiruna, Sweden. At that meeting, the Arctic Council considered applications from 14 candidates for permanent observer status, including six state applicants (China, India, Italy, Japan, Singapore and South Korea), four intergovernmental organizations (the European Union, the International Hydrographic Organization, the OSPAR Commission[3] and the World Meteorological Organization), two environmental organizations (Greenpeace and Oceana), one industry organization (the Association of Oil and Gas Producers) and one other organization (the Association of Polar Early Career Scientists).

The bulk of media coverage given to the meeting focused on the decision to admit the six state applicants as permanent observers, with attention centered in particular on the four Asian shipping nations (China, Japan, Singapore and South Korea). These Asian states' interest in potential Arctic transit routes is hardly surprising given the role that both the Northwest Passage and the Northern Sea Route may some day have in providing commercially viable alternatives that would link Europe, Asia and North America. Reliable Arctic shipping lanes would transfigure commercial relations among the world's trading powers, and China, Japan, Singapore and South Korea have been positioning themselves in anticipation. Furthermore, Arctic Council member states seek overseas buyers for their natural resource exports, and Asian shippers could play a key role in facilitating access to both Asian and non-Asian markets. Enhanced Arctic involvement by Asian shipping states also has the potential to bring new revenues to states such as Iceland that seek to become Arctic transshipment centers and Russia and Canada, which seek to manage coastal portions of Arctic sea lanes.

Not surprisingly, news media in the West reported on the Asian states' bids for permanent observer status (and the Arctic Council's acceptance of the bids) as an intensification of the 'race' for the riches of the Arctic resource frontier. As with so many aspects of the ongoing contestation over Arctic governance, however, the competitive nature of this 'contest' is not so clear. While Arctic and non-Arctic states are indeed strategically positioning themselves, as resource-extraction and transport opportunities open up, the process is essentially cooperative. The goal for members of the Arctic Council, as well as for states that have recently joined the council as permanent observers and for those that will seek to do so in the future, dovetails with the 'resource frontier' imaginary, but it also reinforces the system of state sovereignty. States both inside and outside the Arctic Council are seeking to establish a stable and profitable system for the orderly passage of ships through the region as well as for the removal of resources from the circumpolar North to the more southerly latitudes. States pursuing permanent observer status in the Arctic Council are not seeking to undermine the emergent rule-making process; they merely are attempting to become part of it. And conversely, the Arctic Council does not seek to shut these states out: it just wants to bring them in on its own terms. Differences of opinion certainly exist. However, if there is inter-state competition in the Arctic, all parties have the same goal: the normalization of the North.

This is illustrated by a story *not* reported from the 2013 Kiruna Ministerial Meeting. Media reports focused primarily on the admission of the six new states as permanent observers and secondarily on the deferral of the EU's application, as well as the signing of an agreement to cooperate on marine oil spill preparedness and response. But practically no attention was given to the Arctic Council's failure to approve the other seven organizations' permanent observer status applications. According to participants at the meeting, this was less due to a definitive rejection of the non-governmental and intergovernmental organizations' applications than it was because the ministers simply ran out of time. But that in itself is indicative of the affirmation of the Arctic as a 'normal' intergovernmental space: states – whether full members or observers – 'matter', while the concerns of other entities that claim that their interests are not adequately represented by states are considered only if there is enough time.

Re-imagining the Arctic

Just as the Arctic Council draws on a range of Arctic imaginaries in its efforts to normalize the region within the framework of the modern state system, other parties, who highlight the *exceptional* nature of the Arctic, similarly draw on a range of perspectives. A striking example of this can be seen in an art installation that Greenpeace placed on an ice floe above the Arctic Circle, just before the opening of the 2012 UN General Assembly. The flags that make up the heart represent the 193 member countries of the United Nations, and, according to Greenpeace's website, they 'symbolize an emotional appeal for united global action to protect the Arctic'.

Given that Greenpeace is an environmental non-governmental organization, it is hardly surprising that at the core of the artwork is the 'nature reserve' imaginary – the ideal of the Arctic as a pristine but endangered space that the world community must unite to defend. And it likely goes without saying that Greenpeace's message is directed against the 'resource frontier' imaginary in which the Arctic's riches are to be exploited by those brave pioneers who strive to conquer its forbidding environment, as well as the associated *terra nullius* imaginary under which the conquest of the resource commons is accompanied by an extension of states' political borders to enclose both land and sea.

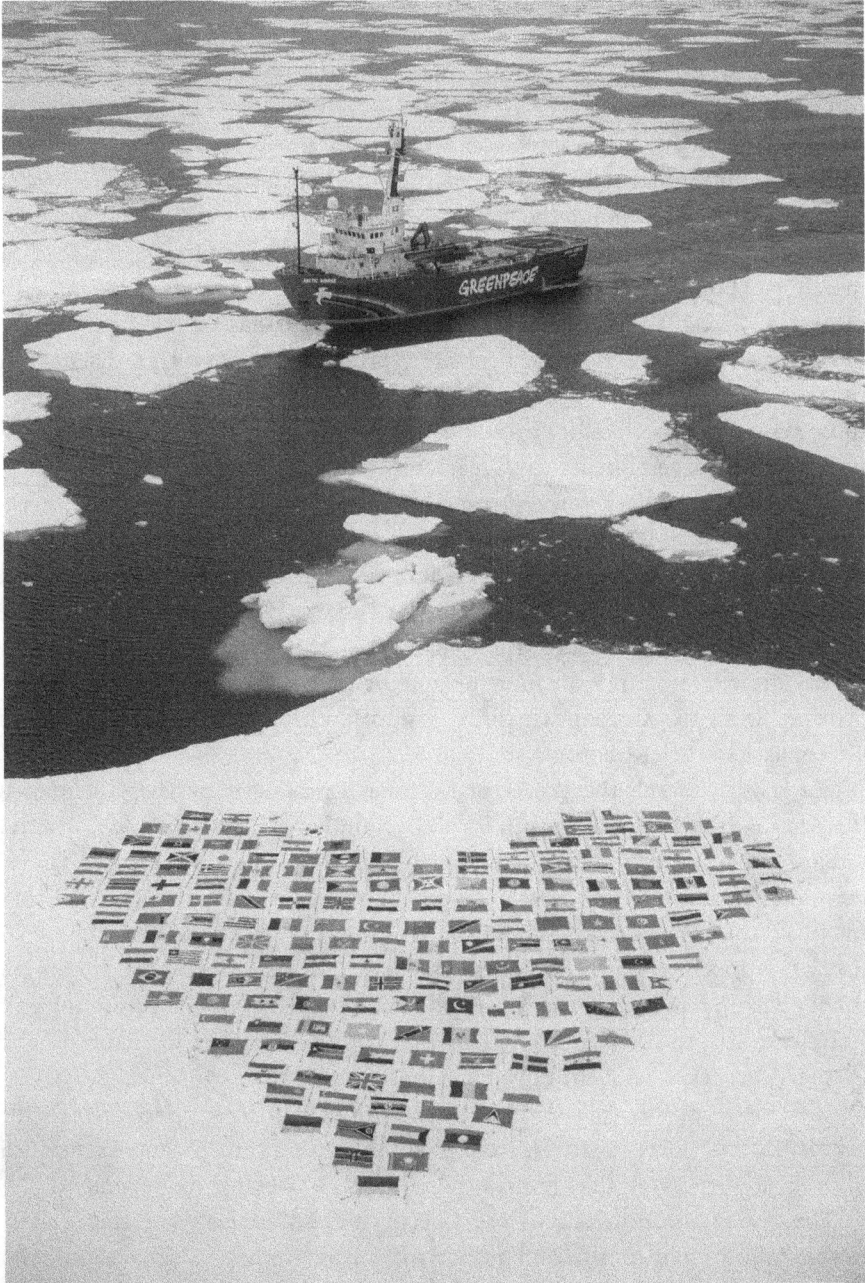

8.2 The crew of the Greenpeace ship MV *Arctic Sunrise* construct a 'heart' with the flags of the 193 country members of the United Nations on an ice floe north of the Arctic Circle.

However, the installation's relationship to the other imaginaries is more complex. Although Greenpeace clearly rejects the idea reiterated at Ilulissat that governance of the region should be allocated to the five Arctic Ocean coastal states, Greenpeace *is* urging action by the *community of states*. In this call to action, Greenpeace is implicitly rejecting the idea that specifically *Arctic* stakeholders have any special claim to the region other than that conferred through the institution of statehood. Interested parties, whether located within or beyond the Arctic, are to filter their views through the policies of national governments (and those governments' votes at UN meetings). This perspective, much like the actions of the Arctic Council with respect to Greenlandic representation and its decisions on new permanent observers at the 2013 Kiruna meeting, reproduces the sovereign distinction between domestic and international affairs, even as it calls for the state system to be tempered by globalist principles. While there may be room in such a vision for new sovereign actors (Greenland could become the 194th flag in the heart, just as it could replace Denmark as the eighth member of the Arctic Council), it would be difficult to accommodate the 'transcendent nationhood' imaginary discussed in Chapter 6. It also removes civil society groups, including, ironically, Greenpeace, from saving the Arctic. Yet while indigenous nationalisms are deemphasized in this vision, the role of civil society groups remains a bit more opaque. Although not present in the heart of state flags, the Greenpeace ship is significantly included at the top of Greenpeace's official photograph.

For most viewers, the power of the image produced by the Greenpeace installation likely lies as much in its background — an Arctic ice floe — as its foreground – a heart made of national flags. Like the condition of Arctic harmony signified by the heart, ice can be fragile and fleeting, a condition made all the more so by climate change. Greenpeace presumably chose to juxtapose the three sets of images – the flags, the heart and the ice floe – to convey the dual message that concerted action is needed by the community of states *both* to protect the Arctic's fragile environment *and* to maintain the fragile peace that is seemingly threatened by conflict. Just as the Arctic Council uses appeals to the region's exceptionalism to support its call for normalization, Greenpeace is using appeals to 'normal' state actors to highlight its message that the Arctic requires exceptional management.

The policy implications of the Greenpeace installation are ambiguous. If the Arctic is so fragile and important an environment, why should its protection be entrusted to the community of states, given that states have little interest in protecting environments beyond their state territory? Why is none of the responsibility being assumed by the region's residents, corporations that are invested in the region, or organizations that have an interest in environmental affairs? Conceivably, all of these actors might have a greater interest in the long-term preservation of the Arctic environment than the extended community of states that is appealed to by Greenpeace.

But of course the Greenpeace installation is not a policy document – it is a work of art. And as such its aim is to raise questions, not provide programmatic directions for solving problems. In particular, as a work of art for the Arctic, the heart of flags needs to raise questions that nourish the generation of *new* imaginaries (or that encourage the reworking of old ones), and, in this respect, the installation is successful. By fusing together three elements that are rarely if ever thought of as existing together – a heart symbolizing love and harmony, flags symbolizing self-interested nation-states, and ice floes symbolizing an imperiled environment – the installation suggests that a hopeful future for the Arctic lies not in conflict between incompatible imaginaries but in their creative combination. Finding the right combination is an agenda shared by proponents of each of the imaginaries discussed in the book, and it is one that they will continue to pursue as they contest the future of the Arctic.

NOTES

Chapter 2 *Terra Nullius*

1. One of the authors of this book (Steinberg) is presently Director of IBRU. However, he had no affiliation with IBRU at the time that the Arctic Region map was produced.

Chapter 3 Frozen Ocean

1. Denmark and Norway are omitted from this discussion because they do not have ice-covered waters adjacent to their main territories. Any ocean laws designed specifically for ice-covered waters would fall within laws for Greenland and Svalbard, respectively.
2. This claim is complicated by a debate over which version of the Treaty is authoritative.

Chapter 4 Indigenous Statehood

1. Despite the apparent parallel with Alaska, which, like Greenland, is discontiguous with the national mainland, there is a key difference in that only 24 percent of Alaska's population is Alaska Native.
2. Nunavut is a partial exception. In Nunavut, a Canadian territorial government structure (the Government of Nunavut), which is not directly connected with any ethnic claim to territory, exists independent of (and, at times, in tension with) Nunavut Tunngavik Incorporated, an ethnically identified organization that is charged with ensuring that the Government of Nunavut honors the conditions of the Nunavut Lands Claim Agreement.

Chapter 6 Transcendent Nationhood

1. It should be noted that there are many disparate indigenous peoples in the 'Arctic' and that there are cultural and political overlaps between them. In this chapter we focus only on the Arctic Inuit.
2. The other five permanent participants are the Aleut International Association, the Arctic Athabaskan Council, the Gwich'in Council International, the Russian Association of Indigenous Peoples of the North, and the Saami Council.

Chapter 8   Normalizing the North

1. The six Arctic Council working groups are Arctic Contaminants Action Program (ACAP), Arctic Monitoring and Assessment Programme (AMAP), Conservation of Arctic Flora and Fauna (CAFF), Emergency Prevention, Preparedness and Response (EPPR), Protection of the Arctic Marine Environment (PAME) and Sustainable Development Working Group (SDWG).
2. The Aleut International Association represents peoples from the United States and Russia, the Arctic Athabaskan Council and the Gwich'in Council International both represent peoples from Canada and the United States, the Inuit Circumpolar Council represents peoples from Canada, Denmark/Greenland, the United States and Russia, and the Saami Council represents peoples from Finland, Norway, Sweden and Russia. The remaining permanent participant, the Russian Association of Indigenous Peoples of the North, represents 43 indigenous peoples from Russia.
3. The OSPAR Commission implements a series of conventions dedicated to environmental protection in the Northeast Atlantic Ocean.

BIBLIOGRAPHIC ESSAY

We conclude with a brief bibliographic essay that directs the reader to articles, books and websites that expand on the themes developed in each chapter of *Contesting the Arctic*. This bibliographic essay is not intended as a comprehensive bibliography of key works on the Arctic but rather as a guide to the specific themes covered in this book.

Chapter 1: Imagining the Arctic

Just as the events of 2007 led concerned Arctic states to issue the Ilulissat Declaration, it also inspired journalists, academics and authors to produce works that addressed 'imminent conflict' in the region. Depending on the perspective of the specific author, these works sought to warn about a coming conflict over the Arctic, rally the populace of a specific nation for that coming conflict, or dismiss the idea that the Arctic is a site where conflict is likely.

Although specifically rejecting the likelihood of all-out war in the Arctic, Roger Howard's *Arctic Gold Rush: The New Race for Tomorrow's Natural Resources* (Continuum, 2009) views the Arctic as a site of unbridled economic and political (and, therefore, potentially military) competition. Barry Zellen in *Arctic Doom, Arctic Boom: The Geopolitics of Climate Change in the Arctic* (Greenwood, 2009), David Fairhall in *Cold Front: Conflict Ahead in Arctic Waters* (I.B.Tauris, 2010) and Richard Sale and Eugene Potapov in *Scramble for the Arctic: Ownership, Exploitation, and Conflict in the Far North* (Francis & Lincoln, 2009) also focus on the Arctic as a site of geopolitical conflict and competition, focusing on economic resources, maritime transport and environmental degradation, respectively. Perhaps the most uncompromising book-length statement of the Arctic as a site of geopolitical competition and potential military conflict is articulated by Scott Romaniuk in *Global Arctic: Sovereignty and the Future of the North* (Berkshire, 2012).

If the Arctic is an arena of emerging competition and potential conflict, then one's own nation must be prepared to exert a presence, whether to preserve the peace or to ensure that one's own national interests are acknowledged and respected by the world community. In Canada, variations on this message are articulated by Ken Coates et al. in *Arctic Front: Defending Canada in the Far North* (Thomas Allen & Son, 2008), Michael Byers in *Who Owns the Arctic?* (Douglas & McIntyre, 2009) and Rob Huebert in *Canadian Arctic Sovereignty and Security in a Transforming Circumpolar World* (Canadian International Council, 2009). In the United States, a similar call to action is issued by Scott Borgerson in 'Arctic Meltdown: The Economic and Security Implications of Global Warming' (*Foreign Affairs*, 2008), and the events of 2007 ultimately led to a wholesale restatement of US Arctic policy ('National Security Presidential Directive 66', online at http://www.fas.org/irp/offdocs/nspd/nspd-66.htm). Norway's perspective on the Arctic as a zone of potential geopolitical conflict (and thus a site that demands attention from national security and economic development establishments) can be found in the Ministry of Foreign Affair's 2011 English-language summary document, *The High North: Visions and Strategies* (http://www.regjeringen.no/en/dep/ud/campaigns/the-high-north/hi gh_north_visions_strategies.html?id=663591). For an analysis of this and earlier Norwegian government initiatives, see Leif Christian Jensen's 'Seduced and Surrounded by Security: A Post-Structuralist Take on Norwegian High North Securitizing Discourse' (*Cooperation and Conflict*, 2013). For the Danish perspective, see *Nar Isen Forsvinder: Danmarks Som Stormagt I Arktis, Olien I Groenland og Kampen om Nordpolen* by Martin Breum (Gyldendal, 2011) and for the Russian perspective, see *Arktika; interesy Rossii i mezhdunarodnyye usloviya ikh realizatsii* by Yuri G. Barsegov et al. (Rossiyskaya Akademiya Nauk, Institut Mirovoy Ekonomiki i Mezhdunarodnykh Otnosheniy, 2002). Even in states far from the Arctic, individuals involved in the security community have been calling on their nations' militaries and diplomatic apparatuses to respond proactively to new opportunities and potential conflicts in the region. Arvind Gupta's 'Geopolitical Implications of Arctic Meltdown' (*Strategic Analysis*, 2009), for instance, urges India not to lose sight of opportunities in the region, and suggests that the best way to protect India's interest in the Arctic would be through the institution of an Antarctica-style treaty system. For overviews of all these positions (and more), see edited volumes by James Kraska (*Arctic Security in an Age of Climate Change*, Cambridge University Press, 2011), Richard Powell and Klaus Dodds (*Polar Geopolitics? Knowledges, Resources and Legal Regimes*, Edward Elgar, 2013), Robert Murray and Anita Dey Nuttall (*International Relations and the Arctic: Understanding Policy and Governance*, Cambria, 2014), and Rolf Tamnes and Kristine Offerdal (*Geopolitics and Security in the Arctic: Regional Dynamics in a Global World*, Routledge, 2014).

Other scholars have rejected the idea that the Arctic is a site of imminent conflict. A relatively optimistic perspective on the Arctic as an emergent sphere of politics is taken by Charles Emmerson in *Future History of the Arctic* (Public Affairs, 2010). Alun Anderson in *After the Ice: Life, Death and Geopolitics in the New Arctic* (Smithsonian, 2009) is even more dismissive of the 'race for the resources' rhetoric that prevails in so much of the popular Arctic geopolitics literature. A further corrective can be found in Shelagh Grant's *Polar Imperative: A History of Arctic Sovereignty in North America* (Douglas & McIntyre, 2010), which, while restricting the scope to Alaska, Canada and Greenland, places the current drive to assert sovereignty (or construct new forms of sovereignty) in the region within a much longer history of state expansion and incorporation/ exclusion of indigenous peoples. The contributors to *Globalization and the Circumpolar North* (University of Alaska Press, 2010) contextualize changes in the region within contemporary processes of globalization.

Although their perspectives are somewhat different from ours, the role of the imagination in southerners' perspectives on the Arctic is explored in a range of works, including Robert McGhee's *The Last Imaginary Place: A Human History of the Arctic World* (University of Chicago Press, 2007) and Barry Lopez' *Arctic Dreams* (Random House, 1986).

The Ilulissat Declaration is available online at http://www.oceanlaw.org/ downloads/arctic/Ilulissat_Declaration.pdf and the United Nations Convention on the Law of the Sea is available online at http://www.un.org/Depts/los/ convention_agreements/texts/unclos/unclos_e.pdf.

Chapter 2: Terra Nullius

The Russian flag-planting, and various reactions to it, are discussed by Klaus Dodds in 'Flag Planting and Finger Pointing: The Law of the Sea, the Arctic and the Political Geographies of the Outer Continental Shelf' (*Political Geography*, 2010), an article that also discusses the processes and performances behind outer continental shelf claims. In his commentary on Dodds' article ('You Are (Not) Here: On the Ambiguity of Flag Planting and Finger Pointing in the Arctic', *Political Geography*, 2010), Philip Steinberg further explores the ways in which sovereignty over Arctic sectors is demonstrated and extends the analysis to cartographic representations of the Arctic. The role of images and imaginaries in the ways in which seabed-mapping is performed and communicated is also discussed by Richard Powell in 'Configuring an "Arctic Commons"' (*Political Geography*, 2008). For a gendered analysis of these performances of sovereignty, see Jason Dittmer et al., 'Have You Heard the One About the Disappearing Ice: Recasting Arctic Geopolitics' (*Political Geography*, 2011). All of these themes are explored further, with specific reference to Canada, by Klaus Dodds in 'Graduated and Paternal Sovereignty: Stephen Harper, Operation Nanook 10, and the Canadian Arctic' (*Environment and Planning D: Society & Space*, 2012).

The history of sectoral claims is discussed with reference to Canada by Robert Dufresne in *Canada's Legal Claims over Arctic Territories and Waters* (Parliamentary Information and Research Service, 2007) and with reference to Russia by Pier Horensma in *The Soviet Arctic* (Routledge, 1991). Further information on the history of sectoral claims, as well as Canada's enclosure of the Arctic archipelago within straight baselines and the declaration of its waters as internal waters, can be found in overviews of Arctic law, including *The Polar Regions and the Development of International Law* by Donald Rothwell (Cambridge University Press, 1996), *Maritime Claims in the Arctic: Canadian and Russian Perspectives* by Erik Franckx (Springer, 1993), *Canada's Arctic Waters in International Law* by Donat Pharand (Cambridge University Press, 2009), and *International Law and the Arctic* by Michael Byers (Cambridge University Press, 2013). The Byers book also contains an extensive discussion of outer continental shelf claims as well as material on the ongoing Beaufort Sea boundary dispute between the United States and Canada and the recently settled Barents Sea dispute between Russia and Norway.

The Atlas of Canada can be found online on the website of Natural Resources Canada (the 'Territories' map is at http://atlas.nrcan.gc.ca/site/english/maps/reference/provincesterritories/northern_territories and the 'North Circumpolar Region' map is at http://atlas.nrcan.gc.ca/site/english/maps/reference/international/north_circumpolar). *Canada's Northern Strategy: Our North, Our Heritage, Our Future* is available online at http://www.northerns trategy.gc.ca/cns/cns-eng.asp. The International Boundaries Research Unit map (and accompanying notes) is available online at https://www.dur.ac.uk/ibru/resources/arctic/ and the US State Department's 'Arctic Region' map is available at http://www.loc.gov/item/2009575046.

For an exceptionally accessible guide to the formulae mandated by UNCLOS for calculating the limits of the outer continental shelf, see 'Defining the Limits of the U.S. Continental Shelf' on the website of the US State Department (http://www.state.gov/e/oes/continentalshelf/index.htm).

Chapter 3: Frozen Ocean

United States of America v. Mario Jaime Escamilla (the T-3 case) and its legal implications are discussed in a number of academic and law journals from the era, including *Arctic* ('State Jurisdiction over Ice Island T-3: The Escamilla Case' by Donat Pharand, 1971), *International and Comparative Law Quarterly* ('International Law and Sea-Ice Jurisdiction in the Arctic Ocean' by Frances

M. Auburn, 1973), *Polar Record* ('Law for Special Environments: Ice Islands and Questions Raised by the T-3 Case' by Daniel Wilkes, 1972; 'Law for Special Environments: Jurisdiction over Polar Activities' by Daniel Wilkes, 1973), and *Western Ontario Law Review* ('Arctic Law and International Ice' by David A. Cruickshank, 1971). The decision of the US Court of Appeals, Fourth Circuit, appears at 467 F.2d 341 (https://bulk.resource.org/courts.gov/c/F2/467/467.F2d.341.71-1575.html).

The classic study of Inuit uses of sea-ice, focusing specifically on Canada, is *Report: Inuit Land Use and Occupancy Project* by Milton Freeman (Indian and Northern Affairs Canada, 1976). More recent studies, which incorporate input from Greenland and Alaska as well as Canada, include *Siku: Knowing Our Ice* by Igor Krupnik et al. (Springer, 2010, as well as the companion interactive atlas website – http://www.sikuatlas.ca) and the articles in the Spring 2011 special issue of *The Canadian Geographer* ('Geographies of Inuit Sea Ice Use'). For a perspective on sea-ice use by non-Inuit indigenous peoples see *Antler on the Sea: The Yup'ik and Chukchi of the Russian Far East* by Anna Kerttula (Cornell University Press, 2000).

Discussions of the status of ice-covered waters in UNCLOS, with particular reference to Article 234, can be found in *The Polar Regions and the Development of International Law* by Donald Rothwell (Cambridge University Press, 1996), *Maritime Claims in the Arctic: Canadian and Russian Perspectives* by Erik Franckx (Springer, 1993), *Canada's Arctic Waters in International Law* by Donat Pharand (Cambridge University Press, 2009), *International Law and the Arctic* by Michael Byers (Cambridge University Press, 2013) and 'Ice Covered Regions in International Law' by Christopher C. Joyner (*Natural Resources Journal*, 1991). Within the United States and Canada, much of the controversy over the application of Article 234 has occurred in the context of the debate over whether the Northwest Passage is an international strait or Canada's internal waters. Key contributions here include, on the US side, works by James Kraska (e.g., 'International Security and International Law in the Northwest Passage' in *Vanderbilt Journal of Transnational Law*, 2009) and, on the Canadian side, works by Michael Byers (e.g., 'Who Controls the Northwest Passage', with Susanne Lalonde, in *Vanderbilt Journal of Transnational Law*, 2009) and Donat Pharand ('The Arctic Waters and the Northwest Passage: A Final Revisit' in *Ocean Development and International Law*, 2007). For a comprehensive review (and proposed compromise), see Philip Steinberg's 'Steering between Scylla and Charybdis: The Northwest Passage as Territorial Sea' (*Ocean Development & International Law*, 2014). Additional works on the interface between security, international law and national development in the circumpolar Arctic can be found in the various contributions to edited volumes by James Kraska (*Arctic Security in an Age of Climate Change*, Cambridge University Press, 2011) and

Myron H. Nordquist et al. (*Changes in the Arctic Environment and the Law of the Sea*, Martinus Nijhoff, 2010).

For comparative perspectives on ways in which various Arctic states have incorporated Article 234 into national legislation and integrated it into national marine resource management frameworks, see 'The Legal Status of Arctic Sea Ice in the United States and Canada' by Betsy Baker and Sarah Mooney (*Polar Geography*, 2013) and 'The Legal Status of the Arctic Sea Ice: A Comparative Study and a Proposal' by Susan B. Boyd (*Canadian Yearbook of International Law*, 1984).

Outside the legal domain, key works on the conceptual problems that arise for southerners encountering frozen ocean that does not easily fit into the binary categories of land or water include those by Adriana Craciun ('The Frozen Ocean' in *Proceedings of the Modern Language Association*, 2010), Urban Wråkberg ('Delineating a Continent of Ice and Snow: Cartographic Claims of Knowledge and Territory in Antarctic in the Early 19th and 20th Centuries' in *Antarctic Challenges*, Swedish Royal Society of Arts and Sciences, 2004) and Kathryn Yusoff ('Visualizing Antarctica as a Place in Time: From the Geological Sublime to "Real Time"' in *Space & Culture*, 2005).

Chapter 4: Indigenous Statehood

For a comprehensive study of Greenland's history from the earliest times, see Finn Gad's comprehensive three-volume *History of Greenland* (C. Hurst & Co., 1970 and McGill-Queens University Press, 1973, 1983). An interesting economic geographical analysis of Greenland when it was still a colony is offered by Herman R. Friis in 'Greenland: A Productive Arctic Colony' (*Economic Geography*, 1937). For a more contemporary analysis of Greenland's economic and cultural situation in the face of greater autonomy, see 'Government, Culture and Sustainability in Greenland: A Microstate with a Hinterland' by Jens Kaalhauge Nielsenby (*Public Organization Review*, 2001). An excellent Danish book that reviews the contemporary social and political realities in Greenland is *Groenland, Maegtig og Afmaegtig* by Marianne Krogh Andersen (Gyldenhal, 2008). The full government report on nuclear activity in Greenland is titled, 'Greenland During the Cold War: Danish and American Security Policy, 1945–68' (Danish Institute of International Affairs, 1997).

The specific juxtaposition between Greenland's potential to become the first Inuit state and the Inuit Circumpolar Council's strategy for greater self-determination is fully explored in Hannes Gerhardt's 'The Inuit and Sovereignty: The Case of the Inuit Circumpolar Conference and Greenland' (*Politik*, 2011). A more focused critique of Inuit political strategies that adopt

Western norms and assumptions is provided by Menno Boldt and Anthony Long in their edited book *The Quest for Justice: Aboriginal Peoples and Aboriginal Rights* (University of Toronto Press, 1985). Another book that more broadly compares the evolving political realities in Greenland and Canada is Natalia Loukacheva's *The Arctic Promise: Legal and Political Autonomy of Greenland and Nunavut* (University of Toronto Press, 2007). The specific issue of establishing a Greenlandic national identity, as opposed to a purely Inuit one, is explored in 'Greenland: Emergence of an Inuit Homeland' by Mark Nuttall *(Minority Rights Group*, 1994) and 'Indigenous Urbanism Revisited: The Case of Greenland' by Frank Sejersen *(Indigenous Affairs*, 2007).

A cooperative, international framework for Arctic governance – the concept of a 'mosaic of cooperation' – is fully expounded in 'Governing the Arctic: From Cold War Theater to Mosaic of Cooperation' by Oran Young *(Global Governance*, 2005*)*. Stephen Krasner explains his theorized conditions of sovereignty in *Problematic Sovereignty: Contested Rules and Political Possibilities* (Columbia University Press, 2001). A Danish book that provides an excellent review of Denmark's vested interest in Greenland and the potentials and pitfalls of finding oil and gas off Greenland's coast is *Nar Isen Forsvinder: Danmarks Som Stormagt I Arktis, Olien I Groenland og Kampen om Nordpolen* by Martin Breum (Gyldendal, 2011). The idea of Greenland being pulled into one or another geopolitical power structure is based on the idea of sovereignty regimes, expounded by John Agnew in 'Sovereignty Regimes: Territoriality and State Authority in Contemporary World Politics' *(Annals of the Association of American Geographers*, 2005).

Lastly, Greenland's relationship to Denmark in terms of identity is well explored from two very different directions via an analysis of Peter Hoeg's novel *Smilla's Sense of Snow*: Prem Poddar and Cheralyn Mealor's 'A Little Country like Ours: Narrating Minority Identity' *(Journal of Postcolonial Writing*, 2008) and Kirsten Thisted's book chapter, 'The Power to Represent, Intertextuality and Discourse in *Smilla's Sense of Snow*', which appeared in *Narrating the Arctic: A Cultural History of Nordic Scientific Practices* (Science History Publications, 2002). Another analysis of the essentializing of Greenlandic identity is offered by Ulrik Pram Gad in 'Post-Colonial Identity in Greenland? When the Empire Dichotomizes Back – Bring Politics Back In' *(Journal of Language & Politics*, 2009).

Chapter 5: Resource Frontier

The frontier and resource frontier concepts were addressed initially by Frederick Jackson Turner's 1893 essay 'The Frontier in American History' (reprinted by Henry Holt & Co., 1921) while Garrett Hardin's 'The Tragedy of the Commons' *(Science*, 1968) has provided a starting point for many studies that interrogate the effects of common property land tenure strategies, a subject that Richard Powell has addressed specifically in relation to the expansion of Arctic oil and gas production in 'Configuring an "Arctic Commons"' (*Political Geography*, 2008).

The linkages between national identities, Arctic nations' resource frontiers and the ways in which national identities are achieved through the 'conquest' and development of these frontiers differ across the Arctic nations, but the relationship between national identity and resource development through-out the region is addressed by Genevieve Ruel in 'The Arctic Show Must Go On' (*International Journal*, 2011) and by Mark Nuttall in *Pipeline Dreams: People, Environment, and the Arctic Energy Frontier* (International Work Group for Indigenous Affairs, 2010). Country-specific works on the role of Arctic resource frontiers in national identity include 'Energy and Identity: Imagining Russia as a Hydrocarbon Superpower' by Stefan Bouzarovski and Mark Bassin (*Annals of the Association of American Geographers*, 2011), 'Geographies of Security and Statehood in Norway's "Battle of the North"' by Berit Kristofferson and Stephen Young (*Geoforum*, 2010), 'Imagining and Governing the Greenlandic Resource Frontier' by Mark Nuttall (*The Polar Journal*, 2012), *Canada and the Idea of North* by Sherrill Grace (McGill-Queen's University Press, 2007) and *Nature's State: Imagining Alaska as the Last Frontier* by Susan Kollin (University of North Carolina Press, 2001).

Roger Howard's *The Arctic Gold Rush: The New Race for Tomorrow's Natural Resources* (Continuum, 2009) is representative of works that identify a 'race for resources' in the North. In addition to the literature that questions the competitiveness of such a 'race', there are also some who question the abundance (or commercial viability) of the North's resources. Critical assessments of the optimistic oil and gas abundance estimates that are typically released by the industry and amplified by national governments include Lars Lindholt and Solvieg Glomsrød's 'The Arctic: No Big Bonanza for the Global Petroleum Industry' (*Energy Economics*, 2012) and Richard Powell's 'Configuring an "Arctic Commons"' (*Political Geography*, 2008). Similarly sobering (although by no means entirely pessimistic) views of the prospects for commercially viable Arctic shipping and fisheries can be found respectively in the *Arctic Marine Shipping Assessment* (Arctic Council, 2009) and 'Current State and Trends in Canadian Arctic Marine Ecosystems I: Primary Production' by Jean-Éric Tremblay et al. (*Climatic Change*, 2012).

Mineral extraction in the Arctic is reviewed by Sharman Haley et al. in 'Observing Trends and Assessing Data for Arctic Mining' (*Polar Geography*, 2011). More focused studies include L. D. Cross's *Treasure Under the Tundra* (Heritage House, 2011), which investigates the growth of the diamond industry in northern Canada, and Colleen Davison and Penelope Hawe's 'All That Glitters: Diamond Mining and Taicho Youth in Behchoko, Northwest Territories' (*Arctic*, 2012), which examines the industry's impact in the Northwest Territories.

A thorough overview of Arctic fisheries is provided in Chapter 13 of the Arctic Council's *Arctic Climate Impact Assessment* (Cambridge University Press, 2006), while existing efforts being made to regulate fisheries are discussed in the *Arctic Fisheries Background Paper* by Erik Molenaar and Hans Corell, written in 2009 for the European Commission's Arctic-Transform project (www.arctic-transform.org/download/FishBP.pdf). Molenaar further discusses what we call 'sovereignty holes' in 'Arctic Fisheries and International Law: Gaps and Options to Address Them' (*Carbon and Climate Law Review*, 2012). Dirk Zellar et al.'s 'Arctic Fisheries Catches in Russia, USA, and Canada: Baselines for Neglected Ecosystems' (*Polar Biology*, 2011) establishes baselines for understanding the rates of fisheries exploitation, while illegal Arctic fisheries are addressed by Magdalena Muir in 'Illegal, Unreported and Unregulated Fishing in the Circumpolar Arctic' (*Arctic*, 2010).

The literature on Arctic oil and gas is extensive. Roman Shumenko's *Arctic Oil and Gas: Development and Concerns* (Nova Science Publishers, 2001) provides a general, if somewhat dated reference, with a focus on Alaska. Aslaug Mikkelsen and Oluf Langhelle's edited volume, *Arctic Oil and Gas: Sustainability at Risk?* (Routledge, 2008) investigates the intersection between states, corporations and indigenous groups with regards to oil and gas development and sustainability, and their focus on corporate social responsibility makes for an interesting comparison with Mark Nuttall's *Pipeline Dreams: People, Environment, and the Arctic Energy Frontier* (International Work Group for Indigenous Affairs, 2010), which is more firmly based in the study of indigenous politics. For an international law perspective that connects the melting of the polar ice cap with changes in international resource development law, see 'Oil and Gas Development in the Arctic: Softening of Ice Demands Hardening of International Law' by Kristin Casper (*Natural Resource Journal*, 2010). Among the many critical works on the oil and gas industry, and its role in modern society and geopolitics, is Matthew Huber's 'Oil, Life, and the Fetishism of Geopolitics' (*Capitalism Nature Socialism*, 2011).

Chapter 6: Transcendent Nationhood

A broad history of Inuit habitation in the Arctic can be found in Robert McGhee's *The Last Imaginary Place: A Human History of the Arctic World* (University of Chicago Press, 2007). The specific role of Knud Rasmussen in pursuing the formulation of an Inuit identity is explored by Philip Lauritzen in *Oil and Amulets: Inuit: A People United at the Top of the World* (Breakwater, 1983). This book also broadly depicts the struggle of the Inuit to establish land claims and self-determination, specifically looking at the creation of the Inuit Circumpolar Conference (later, Inuit Circumpolar Council). Much of Eben Hopson's words, both written and spoken, can be found at the Eben Hopson archive at http://www.ebenhopson.com/.

The works of Justice Thomas Berger are key in the development of land claims and treaty rights for the peoples of northern North America. *Northern Frontier, Northern Homeland* (Douglas & McIntyre, 1971) publishes the findings from a commission that Berger headed to investigate the impacts that the proposed Mackenzie Valley Pipeline would have on indigenous peoples in northern Yukon. In addition to stopping the pipeline, the report went on to have a major impact on southern Canadians' attitudes towards the livelihoods of northern indigenous peoples, which was reflected in the later land claims agreements. Subsequently, Berger was retained by the ICC to chair the Alaska Native Review Commission's ten-year review of the 1971 Alaska Native Claims Settlement Act (ANCSA), which was published as *Village Journey: The Report of the Alaska Native Review Commission* (Hill & Wang, 1985).

For a more recent retrospective on the land claims process, Christopher Alcantara offers a broad review from the Canadian perspective in 'To Treaty or not to Treaty? Aboriginal Peoples and Comprehensive Land Claims Negotiations in Canada' (*Publius: The Journal of Federalism*, 2008). Janet Billson and André Légaré focus specifically on the Nunavut land claims agreement, albeit with very different interpretations, in 'Inuit Dreams, Inuit Realities: Shattering the Bonds of Dependency' (*The American Review of Canadian Studies*, 2001) and 'Canada's Experiment with Aboriginal Self-Determination in Nunavut: From Vision to Illusion' (*International Journal on Minority and Group Rights*, 2008), respectively. The strained relationship between self-determination and sovereignty is analyzed by Jeff Corntassel in 'Indigenous "Sovereignty" and International Law: Revised Strategies for Pursuing "Self-Determination"' (*Human Rights Quarterly*, 1995) while a variety of indigenous perspectives on this relationship, from around the world, are presented in Joanne Barker's edited book, *Sovereignty Matters: Locations of Contestation and Possibility in Indigenous Struggles for Self-Determination* (University of Nebraska Press, 2005). A more specific focus on indigenous peoples and their use of international means to pursue their interests is explored by Timo Koivurova in 'Sovereign States and Self-Determining Peoples: Carving out a Place for Transnational Indigenous Peoples in a World of Sovereign States' (*International Community Law*

Review, 2010), Marjo Lindroth in 'Indigenous-State Relations in the UN: Establishing the Indigenous Forum' (*Polar Record*, 2006) and Austen Parrish in 'Changing Territoriality, Fading Sovereignty, and the Development of Indigenous Rights' (*American Indian Law Review*, 2007).

Jessica Shadian, in particular, emphasizes the emergence of a unique non-state form of sovereignty that the Inuit have achieved, in 'From States to Polities: Re-conceptualizing Sovereignty through Inuit Governance' (*European Journal of International Relations*, 2010). Karena Shaw is another scholar seeking to rethink the relationship between indigenous self-determination and sovereignty, in *Political Theory and Indigeneity: Sovereignty and the Limits of the Political* (Routledge, 2008).

Shadian further explores the use of a constructed Inuit nationalism that is based on claims to land stewardship in 'In Search of an Identity Canada looks North' (*American Review of Canadian Studies*, 2007). This article also analyzes the close and synergistic relationship between the ICC and Canada, which she explores further in *The Politics of Arctic Sovereignty: Oil, Ice, and Inuit Governance* (Routledge, 2014). The relocation of Inuit to establish sovereignty claims in Canada is covered in *Tammarniit (Mistakes): Inuit Relocation in the Eastern Arctic 1939–63* by Frank Tester and Peter Kulchyski (University of British Columbia Press, 1994). Lastly, the struggle faced by Inuit in balancing traditional ways with those of the modern world is explored by Barry Zellen in *Breaking the Ice: From Land Claims to Tribal Sovereignty in the Arctic* (Lexington Books, 2008) and by Shelagh Grant in *Polar Imperative: A History of Arctic Sovereignty in North America* (Douglas & McIntyre, 2010).

The ICC's *A Circumpolar Inuit Declaration on Sovereignty in the Arctic* and its *Circumpolar Inuit Declaration on Resource Development Principles in Inuit Nunaat* can be found at: http://inuit.org/en/about-icc/icc-declarations.html.

Chapter 7: Nature Reserve

Information about the specific Arctic policies of most international ENGOs is readily available online. The WWF's 2009 report on the Arctic Ocean, *International Governance and Regulation of the Marine Arctic*, by Timo Koivurova and Erik Molenaar, is available at http://wwf.panda.org/what_we_do/where_we_work/arctic/publications/?193130/New-Arctic-needs-new-rules-WWF. Greenpeace's *Polar Oceans in Peril and a Planet at Risk* can be found at http://www.greenpeace.org/international/en/publications/reports/polar-oceans-in-peril/, and Oceana's *As Goes the Arctic, So Goes the Planet* is online at http://oceana.org/sites/default/files/reports/Oceana-ArcticPub_2pgspreads2.pdf. Other ENGOs which, although not issuing comprehensive reports, have developed positions on the Arctic environment include Ocean Conservancy (http://www.oceanconservancy.org/our-work/arctic/) and Pacific Environment (http://pacificenvironment.org/section.php?id=181).

The literature on ENGOs and how they cooperate with states in order to pursue their agendas is extensive. Representative of this work is Kim D. Reimann's 'A View from the Top: International Politics, Norms, and the Worldwide Growth of NGOs' (*International Studies Quarterly*, 2006), which details how states have contributed to the rise of NGOs in general, and Kal Raustiala's 'States, NGOs, and International Environmental Institutions' (*International Studies Quarterly*, 1997), which indicates that this is because NGOs, and particularly ENGOs, can actually benefit states. An early book on the topic, Thomas Princen and Matthias Finger's edited volume *Environmental NGOs in World Politics: Linking the Local and the Global* (Routledge, 1994) remains relevant, as does the more recent *NGO Diplomacy: The Influence of Nongovernmental Organizations in International Environmental Organizations*, by Michele Betsill and Elisabeth Corell (MIT, 2007). Thomas Lyon's edited volume, *Good Cop, Bad Cop: Environmental NGOs and their Strategies towards Business* (Routledge, 2010), discusses how different ENGOs choose their battles and what techniques they employ to win these fights.

There is also a number of works that particularly address ENGOs, and, more broadly, the views of environmentalists, in the Arctic. David Standlea's *Oil, Globalization, and the War for the Arctic Refuge* (State University of New York Press, 2006) looks at the issues of ENGOs, development, and business in the context of Alaska's Arctic National Wildlife Refuge (ANWR), while Terre Ryan's 'Creation Stories: Myth, Oil, and the Arctic National Wildlife Refuge' examines how environmentalists, political figures and ENGOs deploy attitudes about 'pristine nature' and wilderness in their debates surrounding Arctic resource extraction (*Journal of Ecocriticism*, 2010). Subhankar Banerjee's *Arctic Voices: Resistance at the Tipping Point* (Seven Stories, 2012) presents a range of attitudes held toward the (Alaskan) Arctic environment by those who live in it, study it, and struggle to protect it.

Antarctic Treaty information can be found at the Antarctic Treaty Secretariat's website, http://www.ats.aq/index_e.htm. This website has copies of all legal documents and their addenda pertaining to the Antarctic Treaty System. Klaus Dodds' *The Antarctic: A Very Short Introduction* (Oxford University Press, 2012) is an excellent source of background information. While most scholars stress the differences between the Arctic and Antarctica, some international law scholars stress that the two regions nevertheless present parallel challenges in environmental governance and that lessons from the two regions can aid regime formation in international law. See, for instance, 'Environmental Protection in the Arctic and Antarctic: Can the

Polar Regimes Learn from Each Other?' by Timo Koivurova (*International Journal of Legal Information*, 2005) and *The Polar Regions and the Development of International Law* by Donald Rothwell (Cambridge University Press, 1996).

Chapter 8: Normalizing the North

As global climate change brings about historic transformations throughout the Arctic region, and as the global community focuses its attention on consequent questions of sovereignty, resource access and potential transit routes, it becomes increasingly difficult to recall the feelings of Cold War trepidation and suspicion that superpower contestation across the Arctic inspired. Ronnie Lipschutz's *Cold War Fantasies: Film, Fiction, and Foreign Policy* (Rowman & Littlefield, 2001) recalls these memories through an historical account of post-World War II international relations together with analyses of the impacts and effects that language and image have on politics, through an examination of 30 Cold War era films.

Just as the Cold War Arctic provided a stage for Hollywood film studios to capitalize on widespread fear and anxiety through a focus on the Arctic, it similarly provided an imaginary for Cold War Arctic science. In 'Frontier Engineering: From the Globe to the Body in the Cold War Arctic' (*The Canadian Geographer*, 2006), Matthew Farish investigates the use of the Arctic as an imaginary for Cold War military science, where civilian organizations with military affiliations promoted Arctic natural and social science research. As the 1980s marked a thaw in Cold War international relations, research specifically on the Arctic's environment emerged as a benign area around which northern antagonists could cooperate. This theme is taken up by Monica Tennberg in *Arctic Environmental Cooperation: A Study in Governmentality* (Ashgate, 2000). In uncovering the antecedents to the formation of the Arctic Council, Tennberg explores how discourses of cooperation, knowledge and development surfaced and were expanded between state and indigenous peoples, around international environmental issues and through different forms and producers of knowledge in the Circumpolar North. In 'The Arctic Council At 10 Years: Retrospect And Prospects' (*University of British Columbia Law Review*, 2007), Timo Koivurova and David Vanderzwaag use the occasion of the Arctic Council's tenth anniversary to assess the work of the council in fulfilling its mandate. In light of further discussions by states over possible changes to the governance regime in the Arctic, Koivurova extended his analysis in 2010 with the follow-up article, 'Limits and Possibilities of the Arctic Council in a Rapidly Changing Scene of Arctic Governance' (*Polar Record*, 2010). Just one year later, the framework for Arctic governance did indeed change, with the passage of the first legally binding agreement negotiated under the auspices of the Arctic Council: the Agreement On Cooperation On Aeronautical And Maritime Search And Rescue in the Arctic (the Arctic Council SAR Agreement). Full text of the Arctic Council SAR Agreement can be found online in English, French and Russian at www.arctic-council.org/index.php/en/about/docum ents/category/20-main-documents-from-nuuk. For an analysis of the Arctic Council SAR Agreement, see Kao Shih-Ming et al.'s 'Adoption of the Arctic Search and Rescue Agreement: A Shift of the Arctic Regime toward a Hard Law Basis?' (*Marine Policy*, 2011).

Notwithstanding its recent expansion into policy making, the bulk of the Arctic Council's work has been through the scientific research produced cooperatively by its six working groups. These scientific reports, available in both full and summary forms, can be found online at http://www.arctic-council.org/index.php/en/about/documents/category/7-working-groups-scientific-reportsassessments. The Cambridge University Press offers the 1,046 page *Arctic Climate Impact Assessment – Scientific Report* (2006) in print form. The full report can also be found online at http://www.acia.uaf.edu/ pages/scientific.html and its synthesis can be accessed at http://www.acia.uaf. edu/pages/overview.html.

E. Carina Keskitalo has done extensive work analyzing the ways in which international political processes influence agenda setting among the Arctic states, and how shifting policies have helped establish the Arctic as a perceived delineable region. She takes up this theme in her book on the origins of the Arctic Council, *Negotiating the Arctic: The Construction of an International Region* (Routledge, 2003) as well as more recent works. In 'Setting the Agenda on the Arctic: Whose Policy Frames the Region?' (*Brown Journal of World Affairs*, 2012), Keskitalo concludes with the sobering assessment that, although international agendas are evolving in the Arctic, the region will remain defined according to political expediency rather than by those who actually live there. Moving from Keskitalo's analysis of how official decisions regarding the Arctic have been made to innovative prescriptions for ways they *ought* to be made, Robert W. Corell's 'Arctic Native Peoples on the Edge' (*Solutions*, 2011) discusses co-management regimes for more effective collaborations between indigenous peoples and states regarding climate change and resource management. More generally, a range of institutional futures for the Arctic has been considered by the Arctic Governance Project (http://www.arcticgovernance.org/). In an analysis that focuses regionally on Canada's High North but which has wider application, Emilie Cameron offers a critique of the assumptions held by scholars, political leaders and the mass media regarding indigenous peoples and climate change, and that guide much of the discussion within both the international policy and national development communities, in 'A Critique of the Vulnerability and Adaptation Approach to the Human Dimensions of Climate Change in the Canadian Arctic' (*Global Environmental Change*, 2012).

In 2011–12, Canada's Walter & Duncan Gordon Foundation sponsored a comprehensive review of the Arctic Council that considered, among other topics, its evolution into a policy-making body and the increasing number of non-Arctic states seeking permanent observer status. A volume that presents papers from the initiative's January 2012 conference, *The Arctic Council: Its Place in the Future of Arctic Governance*, is available online at http://gordonfoundation.ca/publication/530. Piotr Graczyk, one of the contributors to the Gordon Foundation volume, further examines the implications of non-Arctic states increasingly seeking a 'seat at the table', in 'Observers in the Arctic Council: Evolution and Prospects' (*The Yearbook of Polar Law*, 2011).

ILLUSTRATIONS

INDEX

Numbers in *italics* indicate illustrations